DR. TERRI McGINNIS' DOG & CAT GOOD FOOD BOOK

BY TERRI McGINNIS, D.V.M.

ILLUSTRATED BY MARGARET CHOI

TAYLOR & NG SAN FRANCISCO

1977

ISBN 0-912738-09-X

Library of Congress Card #77-6007
Printed in the United States of America

Published by Taylor & Ng
P.O. Box 200
Brisbane, California 94005

First Edition

Distributed by Random House, Inc.
and in Canada by Random House of Canada Ltd.
I.S.B.N. 0-394-73419-X

Photo courtesy of Paul Hinxmann

ABOUT THE AUTHOR

Dr. Terri McGinnis has loved and had a rapport with animals from the time she was a child living in Southern California. Although, or perhaps because, she had few pets while growing up, her ambition was to become a veterinarian. Fortunately, her relatives, teachers, counselors and finally the University of California at Davis concurred. She received her Bachelor of Science degree in 1969 with highest honors and was awarded her Doctor of Veterinary Medicine degree in 1971.

Following her graduation from Davis, Dr. McGinnis moved to the San Francisco Bay area and became engaged in the practice of companion animal medicine. Although she had no plans to become a writer, a wise client made it clear to her that good books about health care for dogs and cats were difficult to find and encouraged her to write one. *The Well Dog Book* and *The Well Cat Book* resulted and their popularity among pet owners in several countries attests to their usefulness.

Dr. McGinnis writes the monthly "Ask the Vet" column for *Family Health* magazine, contributes regularly to San Francisco's KGO-talk radio and has made numerous television appearances. She has come to accept the role of pet health care writer but places most importance on her regular work with animal patients. Dr. McGinnis lives in the Bay area with three dogs, a pet bird and as many flowers, vegetables, wild birds and other friendly creatures as she can lure into her yard.

ABOUT THE ILLUSTRATOR

A native San Franciscan illustrator and designer, Margaret Choi demonstrates her skill, talent, and integrity in *Dr. Terri McGinnis' Dog & Cat Good Food Book.* During the five years she has worked with the innovative team at Taylor & Ng, Margaret Choi has been integrally involved in numerous company projects, including both the design and layout aspects of books, packaging, and promotional materials. A city flat serves as home for Margaret and her 7 year old daughter Kim (born in the Year of the Dog) and as an art studio/office in which she produces her own creative projects.

CONTENTS

To Don,
Thank you for making me
the author I never planned to become.
Your help and encouragement
will always be remembered.

This book could not have been completed without the help of others. Dr. James Morris, Professor of Nutrition and Physiology, University of California, Davis deserves special thanks for reading the manuscript for scientific accuracy. Alan Wood's general supervision, editorial efforts and design suggestions were invaluable. Thanks must also be given to Margaret Choi for her illustrations, Diane Ahlgren for testing the recipes, and to Drs. Jim Milligan, Miles Powers and Tom Reed for making many thoughtful suggestions. All of your efforts were greatly appreciated.

Feeding your dog or cat is the most important thing you do for your pet. Nothing else that you must do daily is more essential to your pet's life. Not grooming or training or even giving love is more important than providing good nutrition. Despite nutrition's importance, however, most pet owners know little about it and those who have tried to study it often have become misinformed.

Books on gourmet cooking for pets have flourished recently, but they offer little accurate information on the nutrition of the dog or cat. Books about animal nutrition are available, but they are often too complicated and boring to be useful to the average pet owner. Some are outdated and some are inaccurate. This book is designed to present up-to-date, correct information about feeding cats and dogs in a pleasing and interesting manner that all pet owners can use. Although this book can serve as an introduction to more complicated information about nutrition, it is not a nutrition textbook for the scientist or breeder who wants to learn how to formulate pet foods. It is a book, however, which gives pet owners facts which can be used to feed their dogs and cats intelligently and economically.

In a time when significant nutrient resources, such as protein, are becoming increasingly scarce, it is important to learn what food sources can be used safely and effectively to feed pets. However, pet feeding is much more than a matter of efficiency and economics since most pet owners value their cats and dogs as friends, aides, and companions who should enjoy their day to day lives including their meals. Commonsense feeding of cats and dogs does not always result in a healthful diet. Since misinformation about the best ways to feed these pets for health and happiness abounds, a practical guide to sound feeding methods is necessary. This book can dispel myths which lead to wasteful and sometimes dangerous feeding practices. It can help you feed your pet in a sensible and nourishing manner. Use it to keep your cat or dog well-fed, healthy and content.

THE WHERE, WHEN & WISE OF FEEDING

Before you even feed your pet his or her first meal there are several basic facts you should know about feeding. Some of the things you read here may seem trivial at first, but they are all important and you will find that most will be useful to you if you will take them into consideration when feeding your pet.

CATS AND DOGS LIKE A FEEDING ROUTINE

Years of experience have shown that most dogs and most cats thrive on routine. Although some pets seem ready, willing and able to eat at any time, even the most adaptable pet is less stressed when served its meals in a familiar place, in a familiar bowl and at a scheduled time without competition from other family pets. This does not mean that every pet must have scheduled mealtimes (see chapter 4), but even pets who have food available at all times need to know that it will be found in the same spot daily and that it will not be pirated by other family pets.

Set aside a specific dining area for your pet and don't change it. An easy to clean corner in the kitchen or laundry area or one near your pet's bed are all good. If you own more than one pet, be sure each has his or her own food bowl (water bowls can usually be shared safely), and that the bowls are separated by a barrier or a distance of at least three feet at feeding time. This simple procedure prevents many fights and can result in remarkable improvements in general body condition for pets who have been intimidated by others at feeding time in the past. If you own both cats and dogs, consider feeding your cats from bowls placed on platforms above your dog's reach or behind a barrier your dog can't squeeze through. Most dogs love to steal the cat's food, and any timid cat can certainly be discouraged from eating by a rambunctious dog. Once the feeding arrangements have been made, don't relax completely. Check things periodically to be sure each pet is getting a fair share.

Some pets seem to thrive when fed by the same person each day. This practice contributes, of course, to the development of a routine about feeding, but it can also cause problems. Pets who become overly dependent on one person for their meals may not be willing to eat on occasions when they must be fed by strangers. To avoid such problems, be sure pets are fed by all family members. In households where this is not possible, the food itself must be the highlight at mealtime. If your pet will not eat without you standing by, measures must be taken to correct the situation.

The importance of having a routine feeding method can become very obvious when you must travel with a cat or dog. Pets which travel frequently to dog or cat shows are notorious for weight loss caused by the stresses imposed by new surroundings, exposure to new people and animals, and the failure to eat when meals are not offered in a familiar manner. Family pets traveling on vacation are subjected to similar stresses. To keep your pet in top condition while on vacation, be sure to take along adequate supplies of a favorite, familiar diet and make feeding time as close to the home routine as possible.

HOW TO CHOOSE A FOOD OR WATER BOWL

Two most important items on any new pet owner's shopping list should be food and water bowls. Many owners feed their dogs or cats from dishes which have previously done duty at their own dinner table. Others, when presented with the array of feeding dishes available at pet stores, select the ones most pleasing to the eye with little regard to the purpose they are to serve. Despite the facts that many unplanned selections turn out to be suitable and that many cats and dogs will find a way to eat from unsuitable containers, there are a few items which, if considered, will help you make an appropriate choice for your pet.

All food and water containers for pets should be made out of materials which are non-toxic, easy to clean, impermeable to grease and germs, and able to withstand the kind of abuse young pets may administer to them as well as the wear and tear of daily use. Containers made of plastic, metal, pottery or glass usually meet these criteria.

Glass or pottery containers are often the most good looking and inexpensive available. They are usually heavy enough to withstand easily being tipped over by the average pet. An attractive, inexpensive earthenware bowl is the old-fashioned beige crock that has been used for years as a food bowl for rabbits. This type of bowl comes in several sizes and can be used for either dogs or cats. If you select a pottery bowl, however, be sure that it has been fired with a non-toxic, lead free glaze.

Plastic bowls come in a variety of shapes and sizes. One can usually be found to meet the needs of any pet, but choose among them carefully. Some become deformed when washed with hot water or crack easily when cold. Others are too soft to be left alone with an active pet. A bored dog may tear one of these to shreds. If pieces are swallowed, they can produce an intestinal obstruction and you can lose more than a food bowl! Some plastic bowls are easily tipped over, but most intended for pets' use are shaped or have weighted bottoms to resist catastrophe.

Metal containers, particularly those made of stainless steel, are probably the best. They withstand hard daily use and vigorous cleaning; they are unbreakable and are inexpensive, particularly when you consider the years of use they provide. When you look for metal food and water bowls be sure to check with a general feed store if you have one near by. A variety of inexpensive metal feed containers made for livestock are also suitable for pet cats and dogs. Do not use galvanized containers, however; they may produce zinc toxicity.

Once you have decided what material is best for your pet's food and water containers, consider what size and shape is most appropriate. Bowls of different shapes have specific purposes. The short-legged dog needs to eat and drink from containers with low sides or it must continually stretch to reach its meal. Bowls with narrow tops and sloping sides are designed to keep long ears out of the food. Cats and small dogs need small dishes, of course, and the converse is true for large dogs. Although most people consider this when choosing food dishes, many fail to provide water containers large enough for big dogs. If you own one of the giant breeds, consider a water *bucket* instead of a water bowl since it is important for water to be available at all times.

Water can also be offered in ways other than from a bowl or bucket. Many cats prefer to drink from slowly dripping faucets. Pet stores carry automatic watering devices for dogs which can be attached to a faucet and provide a drink when licked. For dogs who must be kenneled outside, investigate livestock watering devices which come equipped with warming coils to prevent the water from freezing over in cold weather.

Metal self-feeding bins for pigs are also livestock equipment useful for feeding dry foods to large dogs kenneled outside. Although they are not necessary for daily feeding of most pets, they can be used for feeding when you must be away overnight. Smaller versions can be built from metal or wood for medium-sized dogs. Pet stores carry plastic versions for food and/or water which are ideal for cats or very small dogs.

CLEAN BOWLS AND WHOLESOME FOOD ARE IMPORTANT

All pets' food and water containers and eating areas should be kept scrupulously clean. Frequent washing with hot, soapy water followed by thorough rinsing is, of course, necessary for food and water bowls. There are additional sanitary practices, however, which help keep the feeding area aesthetically pleasing and help prevent disease transmission.

All moist food such as meats, canned pet foods, moistened dry pet foods, leftovers, and liquids such as milk or meat broths should be picked up and refrigerated or discarded if not consumed by your dog or cat within two hours. Many of these foods become unpalatable after sitting and it is wasteful to leave them uneaten. Such foods attract flies, particularly when left outdoors. Those foods remaining palatable enough to be eaten provide an ideal environment for bacterial growth. Foods wholesome when first offered can produce digestive upsets or even severe cases of food poisoning when eaten after remaining at room temperature for a short time.

Commercial soft-moist foods can be left out safely, but many quickly become stale and unappetizing to pets. Dry kibbles or biscuits, however, were designed to remain palatable and wholesome even when left standing at room temperature for days. Feel free to leave

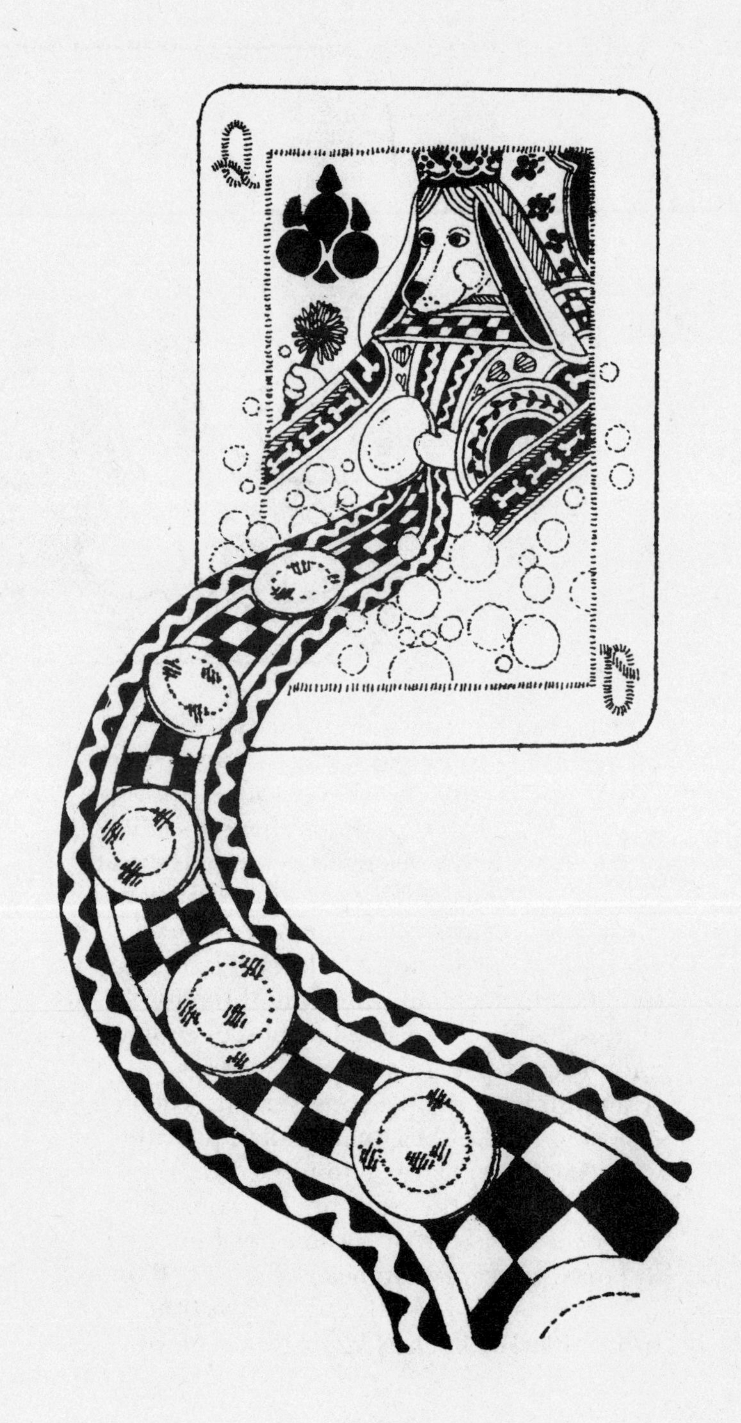

this type of food in your cat's or dog's bowl unless flies or rodents can also reach the food. These pests can carry disease-producing organisms so food contamination by them is to be avoided at all costs. Any dog or cat food left outdoors must, then, be kept in a vermin and fly-resistant container.

Good general sanitation must be followed when feeding cats or dogs because, like people, they are subject to diseases which occur when cleanliness breaks down. Like people also, animals become sick when their food is not wholesome. Unfortunately, many pet owners believe that items *unfit* for human consumption can be fed with impunity to their dogs or cats. This is not the case! Never offer your pet foods which are spoiled or moldy. With the exception of milk or cottage cheese which has just begun to sour, never offer your pet any leftover foods which you would not eat yourself. Water for pets should be fresh and pure. If your home water supply is not safe for you to drink, it is not safe for your pet.

INTERNAL PARASITES CAN BE TRANSMITTED BY FOODS

Trichinosis is a roundworm infection which occurs when larval forms of the worm *Trichinella spiralis,* which encysts in muscle tissue, are eaten. This disease affects humans, pigs, rats and other mammals including dogs and cats. It can cause vomiting, diarrhea and severe muscle stiffness, weakness and pain. You can prevent it in your pet by not feeding raw or undercooked pork in any form. You may want to avoid feeding pork to your pets completely since some pets develop indigestion after eating it. For cats an additional precaution is to try to prevent hunting. Infected rodents can transmit trichinosis if they are eaten by a cat.

Other internal parasites of cats and dogs are also meat borne. Cooking destroys their infective forms so two good general rules to follow are 1) never give your pets uncooked meats to eat and 2) prevent free roaming of dogs and cats who hunt and eat their prey. *Toxoplasmosis,* a protozoan infection capable of causing serious generalized disease in cats and dogs is one of these meat borne diseases. It can, at times, become a public health hazard. Another protozoan, *Nanophyetes salmincola*, is spread via raw salmon. Some dog and cat tapeworms are transmitted in raw, fresh water fish, uncooked meats such as lamb, pork, beef and venison, or prey including rodents and rabbits. The importance of offering wholesome foods and cooked meats, and fish, I think, is clear.

Consider your pet a family member with mealtime needs similar to your own. If you do so, it will become obvious why an eating place and routine free from stress, clean, suitable food bowls, and wholesome foods are so important to a pet. Given a little special consideration, your dog or cat will be more likely to live a long, healthy life.

VITAMINS
B12
FATS
PUP TENT
MINERALS

NUTRITION: ABOUT THE BASICS

Dogs and cats, like people and all other living things, have very specific requirements for nutrients in their diets. Proteins, carbohydrates, fats, vitamins, minerals, and water are all important nutrient substances. A deficiency of one or an imbalance between some can result in health problems. At best a nutritionally deficient pet may become visibly ill in time for the diet to be corrected. At worst signs of deficiency may remain silent until body changes have become irreversible. A young pet may never achieve proper growth and development if fed a diet only slightly inadequate. In order to understand how to feed a pet well, then, it is obvious that a concerned owner must learn something about each class of nutrient and the importance of each one.

PROTEINS

Proteins are essential substances for the body's growth and repair. This fact alone makes them extremely important to nutrition, but they also serve many other essential body roles. For example, nearly all chemical reactions in the body are catalyzed (enhanced) by protein molecules called *enzymes*. Proteins are responsible for the transport of oxygen by the red blood cells and in the muscles. A special protein is also responsible for the transport and storage of iron. *Antibodies* are proteins which help provide the body with protection against viruses, bacteria and other foreign body invaders. Nerve function and motion, the quality we all associate with life are both dependent on protein. When proteins are consumed in excess of the basic daily need, they can also provide the body with energy by being changed to carbohydrate and fat.

All proteins are composed of twenty-three smaller units called *amino acids*. Variation in the kinds and quantities of amino acids present is responsible for the variety of proteins which exist. *Essential* amino acids are protein building blocks which cannot be synthesized in your dog's or cat's body at a rate sufficient to meet the daily need. Because proteins cannot be synthesized in the body from other dietary constituents when the essential amino acids are lacking, they must be supplied by proteins in the diet. To be used optimally, essential amino acids must be supplied in special proportions. Proteins which supply these amino acids in near optimum quantities in a readily available form are said to have a high *biological value*, i.e., they can be used efficiently by the body. Eggs are considered the protein of highest biological quality and are given the value of 100. Other high quality proteins are milk, horse muscle meat and beef liver. Based on biological value, the sum total of several criteria of protein utilization, animal proteins are generally found to be of greatest nutritional value to dogs and cats. Some plant proteins, however, have fairly good biological value, e.g., soybean protein, and are finding an increasing place in dog and cat nutrition as animal sources of protein become increasingly scarce. Cooking increases the availability of some plant proteins (beans) to cats and dogs as does feeding animal protein along with plant proteins, so do not ignore these items as nutrient sources for your pet's diet.

The exact amount of protein required in a pet's diet is influenced by several factors. Protein quality (biological value and digestibility), physiological state of the animal being fed (e.g., growth vs. old age), and caloric density (energy supplied per volume fed) of the diet are all important. Proteins of very low biological value cannot supply enough amino acids to meet a dog's or cat's daily requirements no matter how much of them are eaten. Pet food companies take such factors into consideration when they formulate commercial diets for pets. You must also give such items consideration if you plan to make all your pet's food at home. Protein deficiencies usually occur when pets are fed diets high in carbohydrate containing only plant proteins of low biological value and which are supplemented with animal waste products which are deficient in essential amino acids as protein sources.

The tables on pages 28 and 30 list the protein and other nutrient requirements as they are now accepted for dogs and cats. It is readily apparent from these tables that cats require a diet higher in protein than dogs need. At most life stages this protein need is *at least* twice that of dogs. This is one good reason not to feed cats commercial dog foods. In addition, commercial dog foods may not meet a cat's need for *taurine*, an amino acid essential for normal vision in cats. Cats fed diets deficient in taurine develop retinal atrophy accompanied by blindness.

Depending on the stage in the life cycle, a dog may require up to twenty times the daily protein you do. Cats' requirements may be three to four times those of dogs—up to sixty to eighty times your daily need on a pound for pound basis! It is obvious, then, that as the world's proteins become more dear so may healthful diets for cats and dogs.

EGGS

Eggs are an excellent protein source to be used as a basic part of your dog's or cat's diet or as a dietary supplement. Eggs are a complete protein food containing all the essential amino acids in such ideal proportions that egg protein is the most nutritious known. And, on a pound for pound basis, eggs supply protein to the diet in one of the least expensive ways.

If you plan to use eggs often as a protein source for your pet, be sure to offer them cooked. Cooked eggs are more easily digested and raw egg white contains a substance called *avidin* which binds *biotin* (an essential B vitamin) preventing its absorption from the gut.

MILK

Milk and milk products such as cottage cheese or yogurt are also good sources of protein and provide other nutrients such as calcium, phosphorous and fats as well. Most pets seem to enjoy drinking milk and since milk protein contains an amino acid, lysine, which is not as abundant in cereal protein, the addition of milk to cereals or to pet food products based on cereals is an easy way to improve their protein quality. However, some dogs and cats develop diarrhea when fed any milk products; others may develop diarrhea only when fed large amounts of them.

Diarrhea associated with the ingestion of milk products occurs when *lactose* (milk sugar) is not digested. Failure to digest lactose is due to a deficiency in intestinal *lactase*, the enzyme responsible for breaking down milk sugar. It may also be associated with a lack of intestinal bacteria which break down lactose as they use it for their own nutrition. In either case, undigested lactose attracts water into the intestine causing the diarrhea.

Be sure that milk and milk products provided as sources of protein in the diet are given with care. If loose stools develop when milk is fed, withdraw it from the diet and wait for the stools to return to normal before trying a new milk product. Avoid evaporated milk when introducing milk to a pet unaccustomed to drinking it; its lactose content is twice that of regular cow's milk. In some instances gradual introduction of milk succeeds since its slow introduction allows the intestinal flora to adjust to the new food stuff. In other cases pets which cannot drink milk without developing diarrhea can eat cottage cheese or yogurt which have much lower lactose contents than uncultured milk.

An enzyme powder which can be added to milk before use to break down the lactose has recently become available. If your pet cannot tolerate milk, this product could be of use to you. For more information write to the Sugar Lo Company, P.O. Box 1017, Atlantic City, New Jersey, 08404.

SOYMILK

Soymilk can be an inexpensive milk substitute which may be used as a good protein supplement for a pet's diet. When prepared with the same percentage water as cow's milk, it contains 51% more protein, 15 times as much iron and many B vitamins. Although it is not the good calcium source for pets that cow's milk is, the calcium content is not important when soymilk is used only to improve the protein content of an otherwise well-balanced diet.

Soymilk powder is available in many natural food stores and in Oriental markets. Or you can make your own soymilk. The *Book of Tofu* by William Shurtleff and Akiko Aoyagi, published by Autumn Press, 1975 gives directions for homemade soymilk as well as other soy recipes both you and your pets can eat.

The following is a recipe which can be used as an early morning snack for your pet or as a protein supplement for mixing with a balanced commercial food.

Eggnog For Pets

1 cup cow's or soymilk
1 egg, including shell*
1 teaspoon liver powder (available at health food stores)

Place all ingredients into blender jar and mix until thoroughly blended and foamy.

**Use a soft-boiled egg if you offer this drink more often than an occasional snack or treat.*

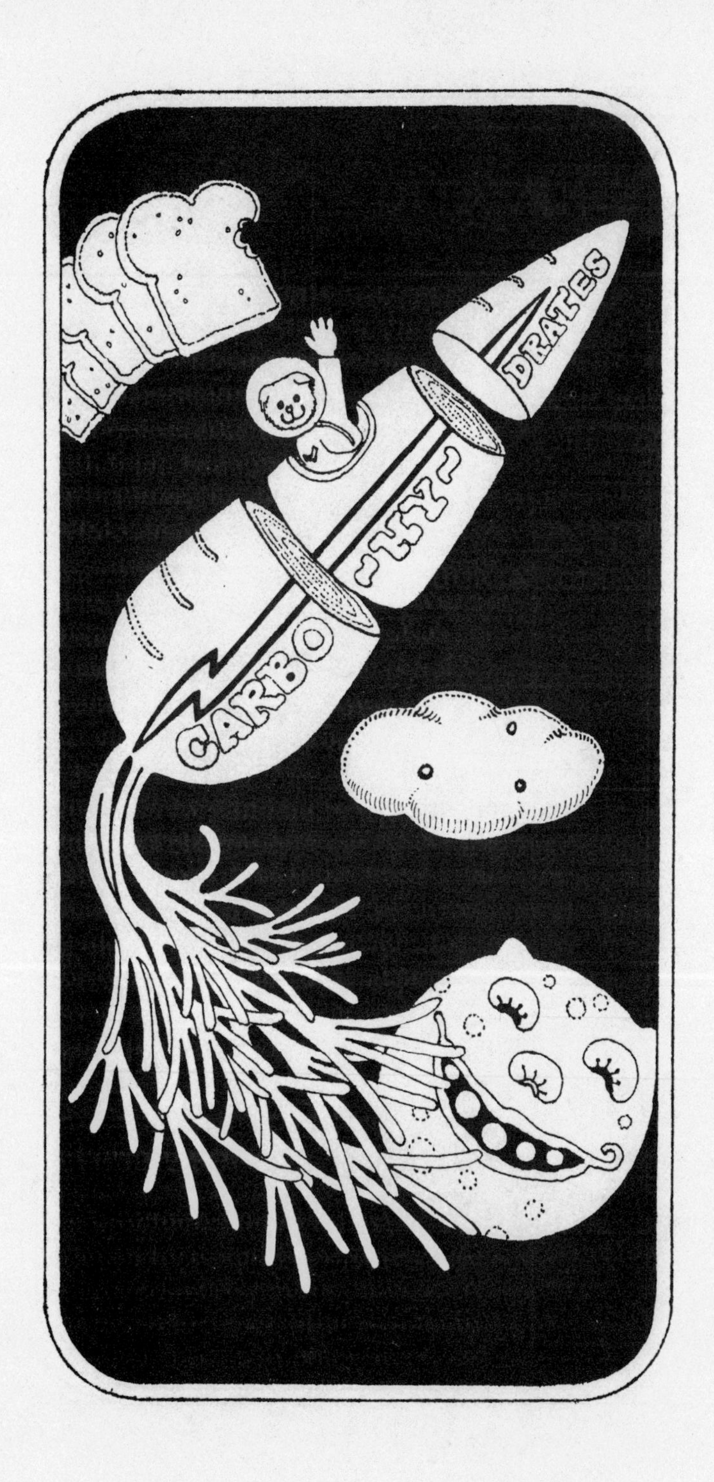

CARBOHYDRATES

Carbohydrates are food substances which are important energy sources to the body. Under normal circumstances the digestible carbohydrates, sugar and starch, are used to maintain the blood glucose level, thereby serving a crucial physiologic role in supplying the brain with energy since it has only very small carbohydrate stores of its own. Although neither the dog nor the cat has been shown to *require* carbohydrate in the diet, both have been found to be able to use starch and sugars very effectively as energy sources. This is an important fact since carbohydrate, used for energy, frees protein (which may also be used for energy) to be used for the important structural purposes in the body that no other foodstuffs can fulfill. In light of the fact that high quality proteins are becoming scarce and expensive today, any relatively inexpensive nutrient which allows us to include less protein in our pets' diets beomes very important indeed.

Carbohydrates are supplied in the diet primarily by plant substances such as grains, vegetables and legumes (peas, beans). Dogs and cats can digest many of these foodstuffs in a raw form. However, some raw starches can induce diarrhea. Grinding and/or cooking helps prevent this problem. Such treatments also free starches and sugars which are sheltered from digestion by *cellulose*, the undigestible carbohydrate which forms plant cell walls.

Although cellulose cannot be digested by your dog or cat, this carbohydrate has its own dietary role. It comprises most of the fibrous part of the diet which stimulates intestinal function and helps prevent constipation. Under normal circumstances the fibrous carbohydrates also control the distribution of water in the intestines preventing diarrhea as well.

Animal nutrition authorities recommend that digestible carbohydrate comprise no more than 33% of a cat's diet measured on a dry weight basis. For dogs no more than 50% to 60% is recommended. These limitations on carbohydrate content are necessary to insure that the diet contain sufficient protein, fat, vitamins and minerals.

FATS

Fats are indispensable to the diet of a cat or dog. They transport the oil soluble vitamins A, D, E and K and supply *linoleic* acid, the unsaturated fatty acid essential to healthy skin and hair. In addition to these unique functions, fats in the diet provide a concentrated source of energy (nine calories per gram as compared to only four for carbohydrates) and have a favorable effect on diet palatability.

The amount of fat your dog or cat requires in the diet depends on the makeup of the fatty acid chains (degree of saturation) and the content of linoleic acid. Deficiency of linoleic acid (or of its metabolites, arachidonic and linolenic acid) results in dry coarse hair and scaling skin which can develop raw areas and is more susceptible to infection. Authorities recommend at least one percent linoleic acid be present in the diet (dry basis) to prevent deficiency in dogs or cats. This amount is usually found in commercial diets for dogs containing no less than five to eight percent fat (dry basis).

Apart from deficiency of essential fatty acid, deficiency in total fat may have an adverse effect on growth. This is especially true for cats. Kittens fed a high fat ration (provided optimal protein is supplied) grow better than those fed diets lower in fats. For this reason, among others (including palatability), recommended diets for cats often contain twenty-five to thirty percent fat. Many commercial canned meats for cats contain this fat level. Many commercial canned fish products, dry foods and soft moist products do not.

HOW TO SUPPLEMENT A DOG'S DIET WITH FAT

Since most commercial diets for dogs contain sufficient levels of unsaturated fatty acids, the idea that a young or old dog with flaky skin needs more fat is probably overworked. Scaly skin and a dry coat can be caused by many distinct disease processes not related to diet. Other dietary deficiencies (e.g., protein deficiency) can also lead to a poor coat. If, however, you think your dog has a fatty acid deficiency, you can supplement the diet with vegetable oil (safflower, corn, soybean or cottonseed oil are good) or lard or bacon fat (not as good). Avoid beef fats or butter fat as supplements since both are low in essential fatty acids. Fats can be added at the rate of one teaspoon to one tablespoon per pound dry food. Canned foods containing 2% to 3.5% fat (as fed) can have fat added at the rate of approximately one tablespoon per one pound fed. Soft moist or canned foods containing more than 6% fat should not have fat added. By increasing the fat content of a dog's diet so that it supplies more than forty percent of the daily calorie requirement, total food consumption can be lowered enough to induce other nutritional deficiencies, so beware. Skin improvement is usually seen one to two months after beginning supplementation if fat deficiency was responsible for the skin problem. A better approach to treating unhealthy skin resulting from dietary deficiencies may be to use a balanced nutritional supplement containing polyunsaturated fatty acids, vitamins A and E, pyridoxine and zinc, or find a new diet and discuss the problem with your veterinarian.

HOW TO SUPPLEMENT A CAT'S DIET WITH FAT

Most commercial dry and soft-moist foods and some canned foods for cats appear to fail to provide fat in optimal quantities. Although the total fat content of many of these products is relatively low, problems do not arise when they provide adequate quantities of essential fatty acid. Products which derive their fat from natural substances high in unsaturated fats such as poultry, fish, or vegetable oil will usually provide adequate linoleic acid even if their total fat content is low. Products which derive all their fat from meat and meat by-products could fail to provide adequate essential fatty acid even if their total fat level was relatively high.

Adding vegetable oil to some cat food products will protect against possible fatty acid deficiency. Dry foods containing 10% fat or less (as fed), soft-moist foods containing 13% fat or less and canned foods containing less than 3% fat can all have vegetable oil (safflower or corn) added at the rate of four tablespoons per pound fed. Canned foods containing more than 4% fat usually do not need fat added. Canned foods containing 5% fat or more should *not* have fat added. If such products are deficient in essential fatty acid, they should be discontinued and a more adequate product fed.

VITAMINS

Vitamins are organic compounds needed by the body in minute amounts for growth, reproduction and to maintain health. They cannot be made in the body so vitamins must be supplied in the diet. They differ from other nutrients in that they are not used by the body for structural purposes or to provide energy. Vitamins are classed into two groups: the oil soluble vitamins A, D, E and K, and the water soluble vitamins also called the vitamin B complex (includes vitamin C).

Although it has been known for many years that cats and dogs, like people, need vitamins, minimum levels of all the essential vitamins have not yet been firmly established. Fortunately, the average well-balanced diet based on natural ingredients (whether made at home or purchased prepared) seems to provide the necessary vitamins in adequate quantities and problems relating to vitamin overconsumption rather than deficiency are seen more often in "well-fed" pets. Therefore, the use of nutritional supplements for healthy dogs and cats known to be fed a balanced diet is not to be recommended.

VITAMIN A

Cats need more than twice as much vitamin A than dogs do to meet their requirements. Since cats cannot convert *beta*-carotene (provitamin A found in green and yellow vegetables) to vitamin A as do dogs, a cat owner must be sure that other sources of fully formed vitamin A are provided in the diet or a deficiency may occur causing skin, eye or reproductive changes. On the other hand, a cat's diet too rich in vitamin A can result in skeletal deformities and crippling.

In order to prevent vitamin A deficiency or excess in a cat's or dog's diet, use a complete and balanced commercial diet (see page 43) and add liver (an organ meat high in vitamin A) only as a supplement to your pet's diet rather than assign it a major role. Feed an adult cat no more than one ounce of beef liver twice a week. One ounce of beef liver per ten pounds body weight each week should be adequate when

used in addition to an already balanced diet for dogs. When necessary for specific health reasons, balanced vitamin-mineral preparations may also be used as dietary supplements to supply vitamin A to pets. Use them for this purpose only upon the recommendation of a reliable veterinarian and follow directions for use carefully.

VITAMIN E

The supposed wonders vitamin E can perform when added to the human diet have received widespread publicity in recent years. This might lead you to wonder what vitamin E supplementation can do for your pet—probably nothing. It is doubtful whether *under normal feeding conditions* vitamin E deficiency (or excess) will occur in dogs and cats.

Unfortunately, however, there have been many cases of vitamin E deficiency in cats. These have resulted when well-meaning but uninformed cat owners have fed their pets excessive quantities of red meat tuna. It has also occasionally followed the feeding of other fish diets, fish oils (e.g., cod liver oil) or large quantities of liver.

Vitamin E plays an important role in the diet by protecting vitamin A from oxidative destruction and preventing rancidity of fats. Within the body it helps protect cells from peroxidation and the destructive effects of free radicals. In vitamin E deficiency in cats, body fat is oxidized and becomes inflamed. This condition is called *pansteatitis* (steatitis) by veterinarians. Affected cats lose their appetite, develop fever and pain accompanied by relectance to move. Untreated cases can eventually end in death.

Pansteatitis should be diagnosed and treated by a veterinarian, but, more importantly, you can prevent its occurence by feeding your cat a sound diet. Be sure your cat's basic diet is a complete and balanced one and avoid frequent feeding of unsupplemented red meat tuna. Any canned tuna fed should be clearly marked—supplemented with vitamin E. Fortunately, reliable cat food manufacturers are aware of the serious problems which can be caused by lack of vitamin E and most now supplement their products. In addition to avoiding fresh tuna or unsupplemented canned tuna, remember to feed liver sparingly and do not use fish oils as routine dietary supplements.

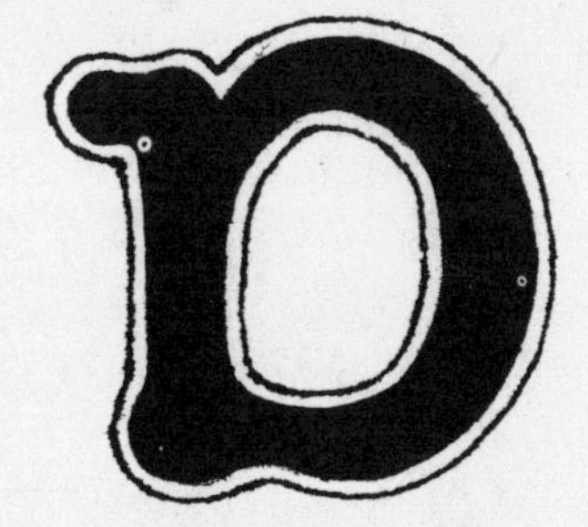

VITAMIN D

Most people know that adequate vitamin D in the diet is necessary for the development of normal teeth and the formation and maintenance of strong bones. Knowing these facts alone, however, has caused many pet owners to imbalance their dogs' and cats' diets and to produce severe metabolic upsets and bone deformities. In the case of nutrition, knowing a few facts can sometimes be more damaging than having no knowledge at all.

Vitamin D fulfils its normal nutritional role by promoting the absorption of calcium from the intestine and by promoting the mineralization of bone. It can only carry out this function, however, when calcium and phosphorus are present in the diet in adequate amounts and in the proper ratio to one another. (For more information about this see page 23). Vitamin D in the presence of too little dietary calcium causes the removal of calcium from the bones followed by skeletal weakness, lameness and sometimes deformity. Insufficient levels of vitamin D interfere with calcium absorption from the gut. Excessive amounts of dietary vitamin D in the presence of adequate calcium and phosphorus may result in excessive mineralization of bone, abnormal teeth, hypertension and calcification of the soft tissues of the body. The delicacy of the interrelationships of calcium, phosphorous and vitamin D is obvious and illustrates the point that unnecessary supplementation of a pet's diet is often a foolish and risky practice. Serious internal changes may be occurring while your pet's external appearance is normal.

Health problems caused by vitamin D deficiency alone are rarely encountered in dogs and cats. Owners have produced hypervitaminosis D in their pets, however, by administering cod liver oil (or other fish oils) in the mistaken belief that it will promote the development of strong bones and a shiny coat or that it will prevent the formation of hairballs.

Cod liver oil should not be used as a routine daily dietary supplement for dogs and cats. One teaspoon of N.F. cod liver oil contains 312 IU of vitamin D. This amount is more than three times the estimated daily requirement for an adult cat and supplies more vitamin D than each pound of dry food should supply to a dog. Regular use, particularly when an already adequate diet is fed, can produce toxicity.

B VITAMINS

The B vitamins are an important group in which several member vitamins are easily destroyed by heating. One B vitamin, *thiamine*, can not only be destroyed by heating but by an enzyme called *thiaminase* which is found in some raw fish and shellfish. Both dogs and cats whose diets have consisted of large quantities of raw fish have developed signs of thiamine deficiency including loss of appetite, vomiting, abnormal reflexes, convulsions and heart failure. Thiamine deficiency is perhaps one of the most readily provoked vitamin deficiencies in dogs and cats.

Problems with thiamine deficiency can be prevented by not feeding raw fish and by carefully selecting commercial foods for your pets. If raw fish must be fed, it should comprise less than ten percent of a dog's or cat's diet. Heat processed commercial foods should have evidence of thiamine and other B vitamin supplementation on the label. This is of particular importance if your pet is a cat since, in most instances where requirements have been established, cats need (on a per kilogram basis) about twice the amounts of B vitamins needed to keep a dog healthy.

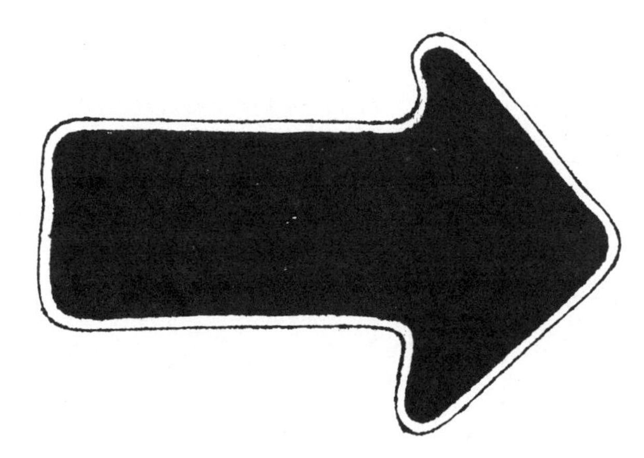

The tables on pages 28 and 30 show currently recommended vitamin allowances for dogs and cats. Use the chart on the following page to learn the role vitamins play in the body and natural food sources of each vitamin.

VITAMINS:

This chart can answer questions you have about the importance of any vitamin to your cat or dog. If your pet's diet contains foods which are natural sources of vitamins, additional supplements are not usually necessary.

VITAMIN	ROLE IN THE BODY	NATURAL FOOD SOURCES
A	*Important for healthy skin, hair, eyes, normal bone growth.*	Dogs can convert provitamin A present in vegetables to vitamin A. Cats cannot. Best sources for both: milk, fat, liver.
D	*Needed for normal bone and tooth formation.*	Formed in the body in response to sunlight. Easy to produce excess in dog or cat. Fish liver oils, egg yolk.
E	*Protects against effects of oxidized or unstable fats in diet. Important to reproduction.*	Present in most foods, e.g., egg yolks, grains, milk fat.
K	*Important to normal blood clotting.*	Synthesized by intestinal bacteria.
THIAMINE (B_1)	*Essential in the metabolism of carbohydrates and protein. Deficiency causes heart and nerve disease.*	Pork, egg yolk, liver, wheat, oats, yeast.
RIBOFLAVIN (B_2)	*Necessary for normal growth, healthy skin and hair.*	Present in most foods. Rich in yeast, organ meats, soy flour, beans, eggs.

VITAMIN	ROLE IN THE BODY	NATURAL FOOD SOURCES
PANTOTHENIC ACID	*Necessary for normal growth, healthy skin and hair.*	Eggs, organ meats, cereals. Also synthesized by gut bacteria.
NIACIN	*Helps maintain the integrity of mouth tissues and the nervous system.*	Rich in natural foods, e.g., brewer's yeast, liver, wheat germ, eggs.
PYRIDOXINE (B_6)	*Essential for the metabolism of protein, normal blood, nerves, and growth.*	Brewer's yeast, egg yolk, wheat and wheat germ, fish, liver, milk, legumes.
FOLIC ACID	*Normal blood; growth and health of the fetus.*	Yeast, organ meats, leafy vegetables.
BIOTIN	*Normal growth, healthy skin and hair.*	Available in many foods. Rich in yeast, liver, milk, molasses. Deficiency produced by excessive feeding of raw egg white.
B_{12}	*Normal red blood cells.*	Muscle meats, milk, liver.
CHOLINE	*Helps prevent fatty infiltration of the liver.*	Liver, eggs, yeast, plant foods.
INOSITOL	*Helps prevent fatty infiltration of the liver.*	Meats, whole grains, plants, yeast.
C	*Necessary to the integrity of body tissues.*	Normally synthesized in sufficient amounts by the dog's and cat's body.

MINERALS

Minerals are inorganic compounds which, like vitamins, are essential nutrients which cannot be manufactured by the body. They are not used for energy but serve very important structural and metabolic purposes in the body. Minerals are found in both the soft and the hard tissues, in body fluids, and are important constituents of many hormones, enzymes and vitamin B_{12}. Major minerals such as calcium and phosphorous interact with one another in delicate relationships which can be easily upset when recommended amounts in the diet are not met or are exceeded. In some instances increased intake of one mineral requires increased intake of other. In other rare cases one mineral may replace another thereby decreasing its requirement in the diet. Small increases in the intake of some minerals such as fluorine or selenium usually required in only trace amounts can produce signs of toxicity. Because of the interrelationships and amounts involved minerals should always be considered as a group and, like vitamins, sources of minerals should be added to your dog's or cat's diet with great care.

CALCIUM AND PHOSPHOROUS

Calcium and phosphorous are the minerals required by the dog and cat in the largest quantities in the diet. Together they form the strong crystalline structure of healthy teeth and bones where 99% of the body's calcium and 80% of the body's phosphorous is found. The smaller quantities of calcium and phosphorous found elsewhere in the body are no less important. Calcium is essential to blood coagulation, muscular contraction, activation of many enzymes, transport of substances through cell walls and the transmission of nerve impulses. Phosphorous has more known functions than any other required mineral. Without it almost every form of energy exchange inside living cells would not occur.

Calcium and phosphorous must be present in your pet's diet in approximately equal quantities. A ratio of 1:1 by weight is adequate for cats. A ratio of 1.2 to 1.4:1 is recommended for dogs. Too little total calcium in the diet or a relative deficiency of calcium caused when the phosphorous intake is too high can cause thinning of the bones, loose teeth, and bone fractures as well as improper growth in the young animal. On the other hand, bone abnormalities have also been seen when excessive calcium is fed. Vitamin D is important to these relationships as well since it affects the body's absorption and transport of calcium. A balanced calcium:phosphorous intake lowers the vitamin D requirement. However, vitamin D added to a diet already low in calcium or imbalanced in calcium and phosphorous results in demineralization of bone, lameness, abnormal conformation and fractures. Such imbalances can also alter the body's requirements for other nutrients such as magnesium.

When dogs and cats are fed a complete and balanced diet, problems with calcium and phosphorous intake are unlikely to arise. However, many pet owners unknowingly upset their animal's calcium:phosphorous balance by feeding unbalanced mineral supplements such as calcium carbonate or bone meal to the diet or by a feeding practice they erroneously believe will improve their dog's or cat's health—the feeding of all meat diets.

Meat is a highly palatable food to most dogs and cats. Few unspoiled pets will refuse a tasty morsel of beef, chicken or lamb, and those who are fed such palatable items routinely, often refuse to eat anything else. If you can afford to feed such a diet to your pet, shouldn't you offer it? If the relatives and ancestors of dogs and cats are carnivores, shouldn't our pets' diets be all meat? The answer to these questions is an emphatic, "No!"

The calcium:phosphorous ratio in muscle meats is about 1:15. Organ meats such as liver, heart and kidney may contain 30 to 50 times as much phosphorous as calcium. In addition, for the dog a diet of beef containing sufficient calories would contain inadequate quantities of vitamins A, D and E, inadequate iron, sodium, potassium, magnesium, and iodine, and excessive amount of fat. Many dogs and some cats become flatulent or develop diarrhea when fed an all meat diet. Such diets may also place an excessive metabolic load on an older pet when kidney failure is present.

Wild carnivores consume the entire carcass of their prey including the bones and the vegetable-filled visera. They also eat insects and nibble on grasses. Clearly, an all meat diet from the meat counter or out of a can is not the answer to providing a nutritious diet for your pet. Use meat only as a supplement to a complete and balanced diet!

Although absolute requirements for all the minerals necessary in a cat's or dog's diet have not yet been established, many important facts about these pets' mineral needs are known. Turn to the chart on page 24 for information about the role minerals play in your pet's body. Current recommendations for mineral intakes for dogs and cats are presented in the tables on pages 28 and 30.

MINERALS:

This chart can answer questions you have about the importance of any mineral to your cat or dog. If your pet's diet contains foods which are natural sources of minerals, additional supplements are not usually necessary.

MINERAL	ROLE IN THE BODY	NATURAL FOOD SOURCES
CALCIUM	*Formation of normal teeth and bones. Important to transmission of nerve impulses and muscle contraction.*	Milk and milk products, molasses, egg shells, bone meal.
PHOSPHOROUS	*Formation of normal teeth and bones. Essential to storage and release of energy in body cells.*	Milk and milk products, beans, eggs, chicken, bone meal, cereal grains.
POTASSIUM	*Necessary to normal growth, body fluid balance and muscle contraction.*	Present in most common foods for cats and dogs.
SODIUM CHLORIDE	*Essential to normal body fluid balance and normal muscle strength.*	Table salt, present in most other common foods for dogs and cats.
MAGNESIUM	*Necessary for normal tooth and bone formation. Important to cellular energy transfer and normal muscle and heart function.*	Present in most common foods for cats and dogs. Rich in soy and other beans, whole wheat and wheat flour.
IRON	*Normal red bloods cells. Important to oxygen transfer in blood and muscles.*	Muscle meat, liver, blood meal, yeast, soy and other beans.
COPPER	*Necessary for normal iron use and for normal bones and hair.*	Present in most common foods for dogs and cats; liver, kidney.

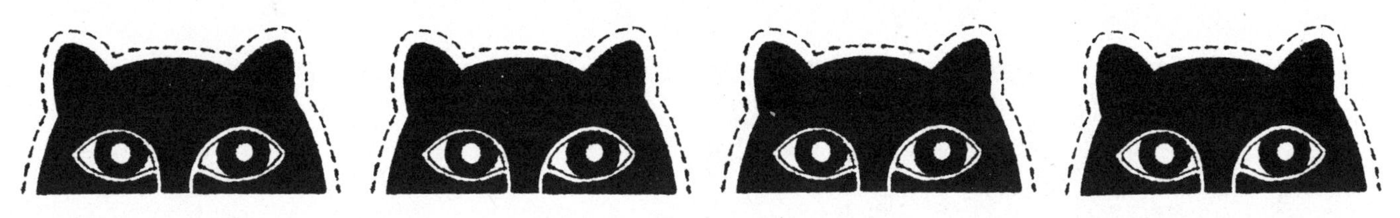

MINERAL	ROLE IN THE BODY	NATURAL FOOD SOURCES
MANGANESE	*Necessary for normal growth, bone formation and reproduction.*	Adequate in most normal food-stuffs; rich in organ meats.
ZINC	*Necessary for normal growth, healthy skin and hair; essential constituent of insulin.*	Muscle and organ meats. (Requirements higher on cereal based vs. meat based diets.)
IODINE	*Necessary for normal thyroid function.*	Sea salt, fish and fishmeal, shell-fish, iodized table salt.
SELENIUM	*Probably essential to normal liver, muscle and reproductive function. Interacts with vitamin E.*	Present as a trace element in most plant and animal tissues.
TRACE ELEMENTS-E.G., COBALT, MOLYBDENUM, FLUORINE, NICKEL, CHROMIUM	*Have diverse functions. Required in minute amounts.*	Adequate quantities probably present in common foodstuffs for cats and dogs.

WATER

Water is perhaps the most important nutrient of all. A dog or cat can go without food for days and lose thirty to forty percent of its body weight without dying, but a water loss of ten to fifteen percent can be fatal. You must, then, be sure your pet gets all the water it requires daily.

Cats and dogs require roughly three-fourths to one ounce water per pound body weight (55 to 65 ml per kg) each day. They obtain this necessary water in the food they eat and the liquids they drink. Water is also a by-product of metabolism. The metabolism of fats by cats is of particular importance in this regard. In fact, the importance of all foods in supplying the water needs of cats is so great that a cat that eats a moist food (75% water) may drink so little as to often lead its owners to believe their pet does not drink at all.

The actual amount of water your dog or cat must drink each day to maintain health is influenced by many factors in addition to diet. Environmental temperature, the amount of exercise taken, the presence of vomiting, diarrhea, or fever are all examples of items which have an effect. Since so many things have an effect on water intake, the best solution to the water problem is to be sure that your pet has access to clean water at all times. Do not give your dog or cat water considered unfit for human consumption, and, if for some reason you are unable to give your pet free water access, be sure to offer water *at least* three times each day.

Following are charts which show the nutrient requirements for dogs and cats.

NUTRIENT REQUIREMENTS FOR DOGS:

This table can help in the evaluation of dog foods. Complete diets should meet or exceed the values in column one. Those that do will provide the required quantities of nutrients listed in column two when consumed in normal amounts. Column two can be used to determine what part of the daily requirement any single food provides when used in conjunction with a table of food composition.

NUTRIENT	Minimum % or Amount Required Per Kilogram Food (Dry Basis)*	Minimum Amount Required Per Kilogram Body Weight Per Day*	COMMENTS
PROTEIN	*18%-22%*	*1.25g-10g*	The amount required varies with the quality (biological value) and digestibility of the protein, physiologic state of the dog, and caloric density of the food. In general, about 20% of the diet's calories should be supplied by protein.
FAT, total (Linoleic acid)	*5%* *(1%)*	*2.2g* *(0.5g)*	If adequate linoleic acid and fat soluble vitamins are provided, protein or carbohydrate can replace fat as a supplier of energy in the diet.
CARBOHYDRATE	*Not more than 50%-65%*	*Not more than 30g*	Not essential in the diet. Necessary to body cells, however, and will be formed from protein or neutral fat if not supplied in the diet.
MINERALS			
Calcium	*1.1%*	*484 mg*	
Phosphorous	*0.9%*	*396 mg*	
Potassium	*0.6%*	*264 mg*	
Sodium chloride	*1.1%*	*484 mg*	
Magnesium	*0.04%*	*17.6 mg*	

**Where a range of values are given, the upper levels are usually necessary to maintain normal growth or lactation.*

NUTRIENT	Minimum % or Amount Required Per Kilogram Food (Dry Basis)*	Minimum Amount Required Per Kilogram Body Weight Per Day*	COMMENTS
MINERALS			
Iron	*60 mg*	*2.64 mg*	
Copper	*7.3 mg*	*0.32 mg*	
Manganese	*5.0 mg*	*0.22 mg*	
Zinc	*50 mg*	*2.2 mg*	
Iodine	*1.54 mg*	*0.068 mg*	
Selenium	*0.11 mg*	*4.84 mg*	
VITAMINS			
D	*500 IU*	*22 IU*	Requirement affected by Ca:P in diet.
E	*50 IU*	*2.2 IU*	
A	*5000 IU*	*220 IU*	
K	*Not required in the diet; normally formed by intestinal bacteria and absorbed.*	*Not required in the diet; normally formed by intestinal bacteria and absorbed.*	May be required during prolonged antibiotic therapy or chronic intestinal disease.
Thiamine (B_1)	*1.0 mg*	*0.044 mg*	
Riboflavin (B_2)	*2.2 mg*	*0.096 mg*	
Pantothenic acid	*10 mg*	*0.44 mg*	
Niacin	*11.4 mg*	*0.50 mg*	
Pyridoxine (B_6)	*1.0 mg*	*0.044 mg*	
Folic acid	*0.18 mg*	*0.008 mg*	
Biotin	*0.10 mg*	*0.004 mg*	Deficiency can be produced by feeding excessive raw egg white.
B_{12}	*0.022 mg*	*0.001 mg*	
Choline	*1,200 mg*	*52 mg*	
Inositol	*Requirement not established*	*Requirement not established*	
C	*Formed in the liver by healthy dogs*	*Formed in the liver by healthy dogs*	Possibly necessary during specific illnesses.

Kronfeld, D.S., ed., Canine Nutrition, University of Pennsylvania, 1972.
Nutrient Requirements of Dogs, ISBN 0-309-02043-0, Committee on Animal Nutrition, National Academy of Sciences–National Research Council, Wash., D.C., 1974.
Wannemacher, R.W., Jr. and McCoy, S., "Determination of Optimal Dietary Protein Requirements of Young and Old Dogs," Journal of Nutrition, 88:66, 1966.

NUTRIENT REQUIREMENTS FOR CATS:

Although all nutrient requirements of cats are not well-established, this table can be used as a guide in evaluating the diet. Complete diets should meet or exceed the values listed in column one. Column two can be used to determine what part of the daily requirement any single food provides when used in conjunction with a table of food composition.

NUTRIENT	Minimum % or Amount Required Per Kilogram Food (Dry Basis)*	Minimum Amount Required Per Cat Per Day*	COMMENTS
PROTEIN	*22%-40%*	*2 g/kg-11 g/kg*	The amount required varies with the quality (biological value) and digestibility of the protein, physiologic state of the cat, and caloric density of the food. In general, about 30% of the diet's calories should be supplied by protein.
FAT, total (Linoleic acid)	*Not less than 15%* *(Not less than 1%)*	*Not less than 7.5 g/kg* *(Not less than 0.5 g/kg)*	Cats may also have a specific requirement for arachidonic acid, a fatty acid available only from sources of animal fat in the diet.
CARBOHYDRATE	*Not more than 33%*	*Not more than 17 g/kg*	Not essential in the diet, however may be used for energy.
MINERALS			
Calcium	*0.77%***	*200-400 mg*	Diets containing more than 2% calcium or 1.6% phosphorous may provoke urinary tract obstruction.
Phosphorous	*0.50%***	*150-400 mg*	
Potassium	*0.58%***	*80-200 mg*	
Sodium chloride	*0.67%***	*1000-1500 mg*	

Kier, A., "Nutrient Requirements of Cats," Southwestern Veterinarian, 27:137-141, 1974.
Nutrient Requirements of Laboratory Animals, ISBN 0-309-02028, National Academy of Sciences–National Research Council, Wash., D.C., 1974.
Personal communication with Dr. James Morris, Professor of Nutrition & Physiology, University of California, Davis, 1977.
Scott, Patricia P., "Nutrition and Disease," Feline Medicine and Surgery, 2nd ed., American Veterinary Publications, Inc., 1975.

NUTRIENT	Minimum % or Amount Required Per Kilogram Food (Dry Basis)*	Minimum Amount Required Per Cat Per Day*	COMMENTS
MINERALS			
Magnesium	*0.04%***	*80-110 mg*	Diets containing more than 0.5% magnesium may promote urinary tract obstruction.
Iron	*40 mg***	*5 mg*	
Copper	*10 mg***	*0.1-0.2 mg*	
Manganese	*56 mg***	*0.2 mg*	
Zinc	*60 mg***	*0.25-0.3 mg*	
Iodine	*2.0 mg***	*0.1-0.4 mg*	
Selenium	*0.5 mg***		
VITAMINS			
D	*1,111 IU*	*50-100 IU*	
A	*28,000 IU*	*1000-3000 IU*	
E	*151 IU*	*4-10 IU*	
K	*Not required in the diet; normally formed by intestinal bacteria and absorbed.*	*Not required in the diet; normally formed by intestinal bacteria and absorbed.*	May be required during prolonged antibiotic therapy or chronic intestinal disease.
Thiamine (B_1)	*4.4 mg*	*0.2-1 mg*	
Riboflavin (B_2)	*4.4 mg*	*0.15-0.4 mg*	
Pantothenic acid	*5.5 mg*	*0.25-1.0 mg*	
Niacin	*44 mg*	*2.6-4 mg*	
Pyridoxine (B_6)	*2.0 mg*	*0.1-0.3 mg*	
Folic acid	*18 mg*	*10 mg*	Formed by intestinal bacteria; unlikely to be deficient in diet based on natural foodstuffs.
Biotin	*10 mg*	*0.1 mg*	Deficiency may be produced by feeding excessive amounts of raw egg white.
B_{12}	*Required; amount not determined*	*Required; amount not determined*	Unlikely to be deficient in diets containing animal protein
Choline	*3,333 mg*	*100-300 mg*	
Inositol	*222 mg*	*10-25 mg*	
C	*Formed in the liver by healthy cats*	*Formed in the liver by healthy cats*	May be necessary during specific illnesses.

**Where a range of values are given, the upper levels are usually necessary to maintain normal growth or lactation.*
***Minerals administered at these dietary levels have maintained normal growth in kittens. Values given here are not meant to represent minimum quantities required.*

FREE WITH PURCHASE OF 13 OZ. BOX OF CHOW MIX...
CHOW MIX
TACO!
CHOW MIX
WOK!
CHOW MIX
BURGER
NEW
PUPP SUPP
PUPP SUPP
PUPP SUPP
CHOW WOW
HI PRO
Special .89
HOT DOG!
MARVEL DOG
WONDER BITS
WONDER BITS
Betty Cocker's
MORE YUM
FURINA
CHOW
CHOW WOW
CONTENTS: 50 CANS

FOOD FACT & FICTION

What should you feed your pet? Since dogs and cats have specific nutritional requirements, balanced diets must be fed for growth and to maintain health. Many commercial foods which claim to be nutritious are found in markets and pet stores. But how can a cat or dog owner be sure such claims are true? What role should commercial products play in the day to day feeding of pets? Should pet owners try to formulate a homemade diet?

HOMEMADE FOODS FOR CATS

The least expensive and most time-saving way to feed a cat for health is to use a good commercial food purchased from a pet shop, grocery store or your veterinarian as a basic diet. However, if you have the interest and the time to cook for your cat and the cost of feeding your pet is not your major concern, a nutritious diet for cats can be prepared at home.

House cats can grow, reproduce and live to old age on a homemade diet which mimics the foods of their wild carnivorous ancestors and relatives. Unfortunately, few cat owners are willing or able to supply a diet of fresh *whole* rodents, birds, insects and plants and such diets cannot be recommended for pet cats from the standpoint of safety (see page 5). An obvious alternative for people who want to give their cats home prepared foods is to offer muscle or organ meats or deboned chicken or fish. These foods, however, tasty to cats, cannot meet their nutrient requirements (see page 23). Instead feeding them causes malnourishment while the cat becomes addicted to a very expensive diet.

Variety and strict attention to diet ingredients are the keys to success when feeding a homemade diet. Since all the nutrient requirements of cats have not been firmly established, a varied diet is the best way to assure that nutrients needed in small but important quantities are not lacking. This is true even when your cat eats some commercial products.

Whole milk and milk products (such as cottage cheese or yogurt), eggs, and meat should comprise three-fourths of a homemade cat diet by weight as fed. The remaining one-fourth of the diet should consist of cooked cereals, whole grains and grain products such as pasta, breads and flours, beans, rice and vegetables. Measure these items' portions before cooking to allow for water absorption and dilution of nutrients. Cats fed a home prepared diet should be given safflower or corn oil daily (two teaspoonsful mixed with the food is adequate for an adult). Complete your cat's diet with a balanced vitamin-mineral supplement purchased from your veterinarian or pet shop and used strictly according to directions. Technically speaking vitamin-mineral supplements are not necessary to an already well-balanced diet; however, the danger of a homemade diet being nutritionally inadequate is so great that a supplement under these circumstances is advisable.

You must be sure that you offer and that your cat eats the milk products and eggs, the vegetables, cereals and other carbohydrates as well as the meats or your home diet will fail. It is easy to fall into the routine of offering a cat the foods he or she adores and to quickly lose the variety that is essential to a nutritious homemade diet. This you must not do.

The meats you offer should be cooked (medium). The cut is not important. However, since cats have trouble eating chunks larger than about ½ inch square, ground meat is probably your best buy. Lamb, beef, and horse muscle meats are all good to feed. (One experiment showed that cats prefer their flavors in that order.) Other good and inexpensive meats are chicken, turkey, rabbit, tongue and spleen. Liver is highly nutritious, but avoid feeding it too frequently since cats easily become addicted to its flavor and vitamin or mineral problems can result (see page 16 and 23). Do not feed lungs or other animal parts you would not eat frequently yourself. They may provide inadequate energy and digestible nutrients for cats.

Cat owners feeding their pets from scratch should keep up on the latest information available about cat nutrition. One publication

which may help you is: *Nutrient Requirements of Laboratory Animals*, National Academy of Sciences, Washington, D.C., second revised edition, 1972 (to be updated 1978).

COOKBOOKS AREN'T NECESSARY

Although cookbooks for cats and dogs are available, you really don't need them. Most pets will eat the necessary foods if they are offered with no more preparation than chopping them into small pieces and cooking or mixing any vegetables offered thoroughly with their meat and eggs. For cats or dogs who pick out the meat goodies and leave their vegetables, cereals, eggs, or cheese you can resort to any good cookbook for ideas. Just use the most simple recipes you can find and leave out all seasonings except salt, garlic or onion.

Here is an example of a simple recipe that will provide a nutritious meal for a dog or cat.

Healthy Meat, Poultry or Fish Loaf For Cats and Dogs

1 pound ground meat or poultry, or cooked and finely chopped poultry or fish

1 cup cooked brown rice

2 eggs, and shells, blended thoroughly until shells are finely broken

½ teaspoon salt

1 cup cooked and chopped green and yellow vegetables (e.g. green beans, carrots, peas)

Combine meat and rice. Stir in blended eggs and shells. Sprinkle salt over vegetables then mix them thoroughly with the other ingredients. Place in a loaf pan and bake at 350°F until done, approximately 30 to 60 minutes.

HOMEMADE FOOD FOR DOGS

Like cats, dogs can be fed nutritious homemade diets if you care to spend the time and money necessary to do it. An old rule that your dog will receive an adequate diet if he or she eats *everything* you do works only if you yourself eat a nutritious diet of basic foods that is extremely high in protein and low (in terms of human needs) in carbohydrate. Although the protein requirements of adult dogs are not much higher than those of people, dogs have very strict protein requirements for growth. Many American diets will not healthfully meet the nutritional needs of growing dogs and the variety of foods people eat can quickly cause digestive upsets in many dogs.

Fortunately, much more printed information is available on the formulation of well balanced homemade diets for dogs than is available for cats. Wise dog owners should make use of it when feeding their pets home prepared foods.

Publications which can help you formulate an adequate diet for your dog are:

Nutrient Requirements of Dogs, National Research Council, National Academy of Sciences, Washington, D.C., 1974.

The Collins Guide to Dog Nutrition, D.R. Collins, Howell Book House, Inc., New York, 1973.

Composition of Foods, Agriculture Handbook No. 8, Bernice K. Watt and Annabel L. Merrill, United States Department of Agriculture, 1963. (This book has information cat owners can also use when a home prepared diet is fed.)

You will probably find that formulating a diet to meet your dog's nutritional needs for growth, adulthood and old age is more difficult and time-consuming than you desire. In that case the safest thing to do is to feed high quality commercial diets. If this is not satisfactory, use some of the following recipes (or ones your veterinarian may be able to supply) to make home prepared meals for your dog. You might also consider mixing some of the following recipes with balanced commercial foods to provide meals half way between home cooking and store bought.

Doggie Delight

½ cup farina (commonly known as Cream of Wheat®), cooked to make 2 cups

1 cup creamed cottage cheese
1 large whole egg, hard cooked
2 tablespoons dry active baker's yeast
3 tablespoons granulated sugar or 1 tablespoon honey
1 tablespoon corn oil or lard

Cook farina according to package directions. Cool. Add remaining ingredients and mix well. This recipe yields about 1¾ pounds food containing approximately 425 calories per pound.*

For growing dogs add to the above:

2 large whole eggs, hard cooked, or
2 ounces canned mackerel, or
1¼ ounce cooked ground beef or lamb or liver

**Originally published in* The Well Dog Book, *Random House/Bookworks, 1974 with thanks to Dr. Morris of Mark Morris Associates, Topeka, Kansas for supplying the basic recipes for these diets.*

My Hero Stew

Cook as a stew:

1 pound ground beef
1 large can stewed tomatoes
6 large potatoes
2 large onions
1 cup macaroni
1 pound dry rice
2 crushed eggshells
2 cups water
Juice of 2 three-ounce cans of yellow beans
Juice of 2 three-ounce cans of green beans
Juice of 2 three-ounce cans of carrots

Add green beans, yellow beans, and carrots, and mix well. This makes ten quarts of food to be fed at the rate of about one quart per forty pounds body weight per day. (This is a restricted protein diet and, as such, may not be suitable for growing puppies.)*

**Originally published in* The Well Dog Book, *Random House/Bookworks, 1974 with thanks to Dr. Morris of Mark Morris Associates, Topeka, Kansas for supplying the basic recipes for these diets.*

Exchange Diet for Dogs

2/3 cup (5 oz. dry) rice, farina, tapioca, or barley
3 oz. medium fat beef, lamb or pork, or chicken, ground or chopped into ½-inch chunks
2 oz. chopped liver
1 teaspoon steamed bone meal or dicalcium phosphate
½ teaspoon iodized salt
2 cups water

Combine all ingredients with water and boil for twenty minutes. Add 1 teaspoonful corn oil if lean meat or chicken is used.

For a high protein diet, substitute one egg for one ounce rice.

Makes enough for a twenty-two pound dog (ten kilograms).*

**Originally proposed by Dr. D.S. Kronfeld at the symposium Diet and Disease in Dogs, U.C. Irvine, November 1975. Published in the article "Canine Diets Analyzed," DVM, pp. 12-14, Vol. 8, No. 2, February 1977.*

KINDS OF COMMERCIAL FOODS

Commercial diets for dogs and cats were originally developed as ways to preserve food, but they soon became ways for pet owners to feed their animals economically and conveniently. In the past many were nutritionally unsound, but today the use of commercial foods gives the pet owner the ability to offer the dog or cat nutrition which usually surpasses that of equally priced home prepared diets. Not only do most commercial diets offer pets good nutrition conveniently and inexpensively, they also help conserve the world's diminishing protein resources by using protein sources not readily available to you or palatable to your pets without special processing. Naturally, feeding commercial pet foods is not foolproof. When an owner walks into a pet shop or market he or she is often confronted by at least ten different brands of foods and more than fifty different products. How can you choose which is best for your pet? Use the information in this section as a guide to help you make this most important decision.

YOU'LL FIND THREE KINDS OF PET FOODS

Commercial foods for dogs and cats are available in three basic forms—canned, semi-moist (soft-moist), and dry. Frozen foods are also available, but not as readily as the other

kinds. Canning pet foods allows the manufacturer to use fresh, wet ingredients, cook them and preserve them so you can feed them conveniently at home. In general canned foods are the most expensive products for feeding cats and dogs (you pay for 65-78% water) and they vary most in content and quality. Some canned foods contain only meat and/or fish and meat (fish) by-products. These products fed alone cannot provide balanced nourishment for a dog or cat (see page 23). Other canned foods contain meat (and/or fish, and/or by-products) and just enough vitamin-mineral supplement to make them a complete and balanced diet for a cat or dog. The last group of canned foods includes a great variety of products which use meat or fish as a base but also add eggs, vegetables, cereals or flours to balance the nutrition, improve the flavor or "fill them out". These products are commonly called "canned mixed diets" or "canned ration type" foods. Some of these foods will provide complete nourishment for your pet. Others cannot and you must be very selective when choosing among them.

Canned foods are usually highly palatable to cats and dogs. Although this allows you, if you are careful, to select and feed products which provide complete nutrition and are eaten readily, feeding them freely can easily lead to overeating. In pets not accustomed to canned foods, digestive upsets, such as diarrhea and gas, often follow such overindulgence. When overeating becomes a routine, obesity soon results (see page 77).

Semi-moist foods are a technological triumph. The processing these products undergo allows foods with a moisture content that would usually cause them to spoil at room temperature to be stored unrefrigerated and to be fed conveniently from a package. In general, all soft-moist products for dogs and for cats are relatively uniform in ingredients and are designed to provide complete nutrition. They are usually two to three times as expensive to feed as dry foods, but less expensive than canned foods when water content is considered. Most pets find them very palatable; when semi-moist products are offered, mealtime is usually over quickly with a minimum of effort and little or no clean up.

Manufacturers are able to produce these soft tidbits in a package by mixing and cooking the basic food ingredients (soybean meal, meat and/or meat by-products, fat, vitamins and minerals) with artificial flavors and colors, large quantities of humectants and sugar (dog foods) or phosphoric acid (cat foods). Humectants (such as sorbitol or propylene glycol) hold water keeping the foods soft and helping to prevent spoilage. The sugars (sucrose most often) or acids also act as preservatives since bacteria and molds have difficulty growing in a high sugar or acid environment. Unfortunately, humectants, preservatives, large quantities of sugars and artificial flavors and colors do not resemble the natural foods dogs' and cats' bodies evolved to metabolize. Although sorbitol, propylene glycol and sugar can all be used for energy, some pets develop digestive upsets when fed soft-moist products containing such ingredients. Until the long term effects of feeding such foods are apparent, I think it is best to use semi-moist products only as a treat or to coax a lagging appetite, not as the regular diet.

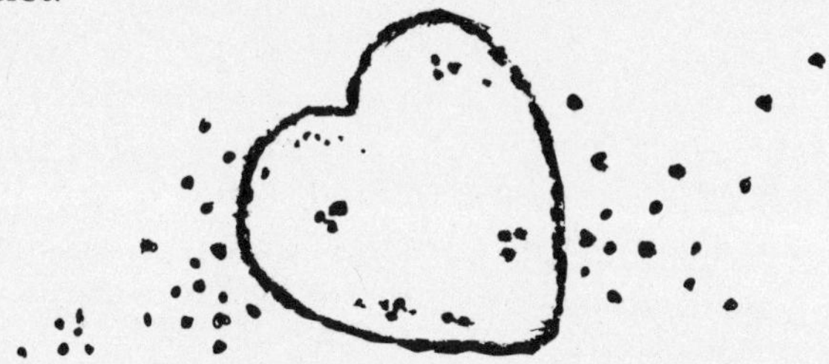

Dry dog and cat foods combine several characteristics desirable in pet foods. They are easy to transport and store, convenient to feed, resist spoilage when left in the food bowl, and are the least expensive commercial pet foods. All complete dry foods are similar in the sense that they are based on a combination of cereals, grains and flours, meat or fish products, sometimes dairy products, and vitamins and minerals which are processed to produce a dry, crunchy product containing about eight to twelve percent water. However, at this point their similarity stops. Dry foods for dogs come in many shapes, sizes and colors—biscuits, kibbles (broken up biscuits), meals, nuggets and pellets. Some products make a "gravy" when water is added, others do not. There is only slightly less variety among cat foods, the uniformity which is present occurs because cats seem to have trouble picking up pieces larger than about ½ inch. Most major brands of dry foods for cats and dogs are designed to provide a complete and balanced diet for your pet, but some brands are not. As with other foods, you must select a dry product carefully to insure proper nutrition.

Most dogs and cats find dry foods palatable, especially if they have been fed them from an early age. Pets which have been fed "people foods" and meaty canned foods while young may refuse dry foods at first but this problem can usually be solved relatively easily (see page 61). Free choice feeding of dry foods can usually be allowed without the danger of obesity, and dry foods help promote dental health in cats and dogs.

Some authorities question the protein content of dry dog foods. Any possible difficulties can usually be avoided by selecting complete and balanced dry foods with protein contents greater than 18% dry matter for adults and greater than 28% dry matter for puppies. For additional insurance, particularly for growing, working or nursing dogs, you can give your pet relatively inexpensive supplements such as powdered dry or fresh milk, eggs, or cottage cheese or use the more expensive protein supplementation of fresh or canned meats. Even with protein supplements added, dry foods give you the best dollar value in nutrition.

Recently unfavorable comments have been made in the press about the advisability of feeding cats dry foods. Some individuals have suggested that dry foods may induce urinary tract disease in cats. Although many veterinarians suggest changing the diet of cats stricken with urinary tract disease, this does not mean that they feel the diet *caused* the problem.

There is no generally accepted scientific evidence that dry foods are the primary cause of bladder inflammation and urinary tract obstruction in cats. As long as you make sure that adequate water supplies are always available to your cat, it is safe to feed dry foods to your pet unless health problems which require special dietary restrictions already exist.

The primary draw back of dry foods lies in the fat content. The physical characteristics of the standard dry foods prevent incorporation of fat at a level much greater than about 10%. During storage some of this fat is lost by exudation or deterioration so it is of no use to the cat or dog. If fat is not added to some dry cat and dog foods, a deficiency of essential fatty acids may occur. Fortunately any problems associated with fat content in dry foods are usually easily corrected. (see page 14).

HOW TO CHOOSE A COMMERCIAL FOOD

Commercial pet foods and their labels fall under the regulation of the Federal Food and Drug Administration, the Federal Trade Comission and the United States Department of Agriculture as well as individual state regulatory agencies. These groups generally insure that pet foods are pure and wholesome, contain no harmful substances and are truthfully labeled. These policies are enforced in many states by random sampling and testing of pet foods off supermarket shelves. Unfortunately these groups do not yet require enough information to be present on pet food labels to enable you to compare one product *directly* with another. There is, however, enough regulation that information found on the products sold at your local supermarket or pet store can help you make an intelligent choice of foods for your pets based on a process of elimination. And as time passes, more and more useful information will become printed on dog and cat food packages as a result of increased consumer interest and the efforts of the Association of American Feed Control Official (AAFCO). This group consists of animal feed control officials from throughout the United States and Canada who enforce laws regulating the production of foods for animals and develop laws and regulations which promote effective pet foods and useful and truthful pet food labels.

The selection of a commercial diet for your cat or dog must be done in a logical manner. There are six items found on all pet food labels that can be of most help to you. They are 1) the nutritional statement, 2) the list of ingredients, 3) the guaranteed analysis, 4) the net weight, 5) the cost, and 6) the manufacturers name and address. Examine each item in turn, then use all the information you have gathered to make your final food choice. Even if you use more than one product to feed your pet, care should

be given to each food selected. Feeding even a single poor quality product can significantly detract from the all around good nutrition of your dog's or cat's diet.

HOW TO USE THE NUTRITIONAL STATEMENT

The most important piece of information on a pet food label for the average pet owner is the *nutritional statement*. Increasingly uniform regulation of pet foods virtually guarantees accurate label statements about pet food ingredients, but the proliferation of dog and cat food products for various purposes can make foods compounded of the same ingredients inadequate for one pet's needs while desireable for another's. Use the nutritional statement to determine whether any cat or dog food should be given further consideration as part or all of your pet's diet.

Dog or cat foods which state on the label that they are *complete*, *perfect*, *scientific* or *balanced* rations or indirectly imply that they are by statements such as "All your pet ever needs" must contain sufficient ingredients to meet the nutrient requirements of cats or dogs at *all* life stages. Companies which make such statements on their product labels must include appropriate ingredients in quantities sufficient to meet the nutrient requirements established by the National Research Council of the National Academy of Sciences (or an equivalent nutrition authority) or must have undergone testing which proves they will support proper growth, adult activity and reproduction when fed *without supplementation*. Foods which say they provide complete and balanced nutrition for your dog or cat (or make an equivalent statement) are the *only* products which can be fed without additional effort being made to balance your pet's diet. Products without statements attesting to their nutritional completeness are not guaranteed to provide total nutrition for your pet and must not constitute your cat's or dog's complete diet if you are to be sure proper nutrition is being provided. They can, however, be used as part of a varied diet.

Some commercial foods for dogs and cats are designed to provide complete and balanced nutrition *only at particular life stages*. Companies which produce these specialty foods are allowed to label these products which are suitable for limited purposes as complete or balanced diets only in conjunction with a statement of that limited purpose. Unsuspecting cat or dog owners must beware! Such products, while perfectly good for their intended use, can cause trouble when fed to pets for which they were not intended. Foods designed for adult pets may not provide adequate nutrition for growth. Foods for pregnant or nursing mothers will provide too much nutrition for the old, fat female. Trim working dogs fed weight reduction diets will have to consume huge amounts of them in an attempt to obtain enough energy to meet their daily needs.

Read every commercial product's nutrition statement carefully and use it to eliminate from consideration foods which are not suited for your pet's diet. If you are not sure what the nutrition statement means, ask your veterinarian for help.

HOW TO USE THE LIST OF INGREDIENTS

Once you have narrowed your choice of foods down to products that fall into a class your pet can use, the list of ingredients on the product labels can further aid you in your selection. By law, pet food ingredients must be listed in order of their respective predominance by weight in the product. Although the ideal ingredients list would tell you the exact quantity of each ingredient present and allow commercial pet foods to be compared directly with one another, the current method of listing ingredients does allow you to eliminate some pet foods from further consideration.

The best foods for feeding cats and dogs contain a source of animal protein found among the first or second (canned foods) to third or fourth (soft-moist or dry foods) ingredients. Balanced diets for dogs and cats can be produced from vegetable products alone but meat protein not only improves the palatability of dog and cat foods, but also the utilization of proteins from vegetable sources. Proteins in foods where animal tissues fall low in the list of ingredients may not be as well utilized by your dog or cat as those in products containing more animal protein. You may pay the same price for both products, but your pet may receive less nutrition from the product whose company skimps on animal protein. (For more information about protein utilization see page 49.)

Other foods you should seriously consider eliminating from your deliberation are those containing artificial colors or food additives such as sodium nitrite. Artificial colors and other additives may be included in pet foods if they are listed in the Federal Government's "Generally Recognized As Safe List" (GRAS list). Additives not on the GRAS list may also be included depending on the circumstance; however, feed control officials *may* require the maker to submit evidence attesting to the safety of the pet food containing such additives. Unfortunately in years past food additives reached the GRAS list without extensive testing and additives once considered safe have only recently been removed from the list after being implicated in health problems. (Red dye No. 2 is a good example.) Except for substances added to balance the vitamin and mineral content of pet foods, most additives are included to impart characteristics desirable to pet owners, not pets! Cats and dogs who have little, if any, color vision don't care if their food looks like yellow bits of cheese and red bits of meat in a gravy. They don't require that cooked meat in a can retain the pink blush that indicates freshness to most owners, a blush that can only be maintained with preservatives. Since our pets place most importance on the texture, taste and odor of their food, we as pet owners should learn to rely less on appearances. By refusing to purchase pet foods with unnecessary additives and writing to the food companies when you have a complaint, you can encourage the most wholesome foods to be produced.

It is important not to confuse substances added to pet foods to correct or improve the vitamin-mineral content or to prevent fat rancidity with undesirable additives. Heat processing and storage can destroy necessary vitamins and fats present in natural ingredients and certain foods used in the preparation of cat and dog rations are initially deficient in important vitamins and minerals (e.g., meats). Common safe vitamin-mineral supplements used in commercial cat and dog foods are potassium chloride, calcium carbonate, choline chloride, iron oxide, potassium iodide, cobalt carbonate, calcium pantothenate, manganous oxide, zinc oxide, ferrous sulfate, copper oxide, menadione sodium bisulfite, thiamine, biotin and folic acid. The tocopherols (vitamin E), BHA (butylated hydroxyanisole) and BHT (butylated hydroxytoluene) are used most often to preserve fat. You may obtain a copy of the Official Publication of the Association of American Feed Control Officials, Inc., by writing to their treasurer. (A current address can be obtained from your local reference librarian.) This publication contains official and tentative definitions of animal food ingredients and will help you understand what is meant by meat by-products, wheat middlings and the names of other pet food ingredients. If you don't need such detail, when you find an ingredient you question you can ask your veterinarian for an opinion or for a less detailed reference that may help you.

HOW TO USE THE GUARANTEED ANALYSIS

The *guaranteed analysis* must appear on the labels of all commercial cat and dog foods. The information which is required to appear must be listed in the following order:

crude protein (minimum amount)
expressed in percent
crude fat (minimum amount)
expressed in percent
crude fiber (maximum amount)
expressed in percent
moisture (maximum amount)
expressed in percent

Additional guarantees, such as for the minimum amount of calcium or phosphorous present, may follow the above information but are not required by law.

The guaranteed analysis can provide you as a pet owner with a lot of information if you keep some important facts in mind. Protein and fat content (the more expensive ingredients) usually approach the minimum amounts listed and the less expensive ingredients approach the maximums. This is not inherently bad; use it to your advantage. For example, if the minimum protein content in a food you are considering just barely exceeds the minimum amount recommended for your pet, eliminate it immediately from your deliberations since the product is unlikely to contain much more protein.

The guaranteed analysis tells you nothing about the *digestibility* of the dog or cat food. No matter what a product's guaranteed analysis states, if the ingredients are not highly digestible they are not available for use by your pet. Fortunately, reputable manufacturers formulate their commercial diets from high quality, wholesome foods and food products and the list of ingredients can help you eliminate products made from the poorest ingredients. For detailed information about the digestibility of a product you will have to write the manufacturer.

HOW TO USE THE MANUFACTURER'S ADDRESS

The name and address of the manufacturer, packer or distributor of a dog or cat food is also information required by law on every pet food label. If the address is not listed, it is present in the local telephone directory of the city shown on the label. This can be an important tool for you. Use it to request more detailed information about the company's foods. You can also use it to direct any complaints you have to the manufacturer. (Remember to send the food's identifying code number whenever you have a complaint.)

On the following page you will find a pet food evaluation chart.

Use the guaranteed analysis along with the price of the food and package weight to fill in the pet food evaluation chart. Information you fill into the columns on this chart will help you make a final choice among the pet foods you are considering.

See page 48-49 for instructions on how to fill in the Pet Food Evaluation Chart.

PET FOOD EVALUATION CHART

BRAND NAME	PROTEIN		FAT		FIBER		% MOIST-URE	% DRY MAT-TER	PRICE	PRICE	PRICE	PRICE
	% as fed	*% dry weight*	*% as fed*	*% dry weight*	*% as fed*	*% dry weight*			*per pkg. weight*	*per lb. sold*	*per lb. dry matter*	*per lb. protein*
COLUMN NO. 1	2	3	4	5	6	7	8	9	10	11	12	13

Photocopy this page and fill in the columns to help you decide which food to choose for your pet.

HOW TO FILL IN THE EVALUATION CHART:

1. Use the pet food evaluation chart to compare only complete and balanced commercial foods which have withstood your initial criteria for elimination. Rations not labeled complete and balanced are designed only to be diet supplements or treats for your pet.

2. Fill in columns 1, 2, 4, 6, 8 and 10 from the information found on the package.

3. Calculate the percent dry matter, column 9.

100% − % Moisture = % Total Dry Matter

Since food's energy and nutrition is present only in its dry matter, foods cannot be compared adequately with one another unless differences due to water content are eliminated.

4. Fill in columns 3, 5 and 7.

$$\frac{\textbf{Guaranteed \% of Nutrient as Fed}}{\textbf{\% Dry matter}} \times \textbf{100}$$

= % Nutrient
Present Dry Basis

PROTEIN*

Products containing less than 34% protein dry matter are not suitable for feeding kittens or nursing mothers.

Products containing less than 30% protein dry matter should not be fed to adult cats.

Products containing less than 28% protein dry matter are not suitable for feeding puppies.

Products containing less than 18% protein dry matter should not be fed to adults or puppies (unless specifically recommended by your veterinarian.)

**Assumes 80% utilization. Protein levels may need adjustment up or down depending on the caloric density of the diet.*

FAT:

Products containing less than 15% fat dry matter should not be fed to cats without fat supplementation. Products containing 25% to 30% fat dry matter do not require supplementation.

Products containing less than 5% fat dry matter should not be fed to dogs without fat supplementation. Products containing more than 8% fat dry matter do not require supplementation.

FIBER:

When choosing between products that are equal in other respects, eliminate the product higher in fiber content. Since dogs and cats require little fiber in their diets, it is not economical to pay for this undigestible material. (Speciality diets for weight reduction are high in fiber to give pets a feeling of fullness.)

5. Fill in column 11.

$$\frac{\textbf{Price}}{\textbf{Product Weight In Ounces}} \times 16$$

= Price Per Pound Sold

6. Fill in column 12.

$$\frac{\textbf{Price Per Pound Sold}}{\textbf{\% Dry Matter}} \times 100$$

= Price Per Pound Dry Matter

When choosing between products equal in all other respects, use this column to determine which one is most economical.

7. Fill in column 13.

$$\frac{\textbf{Price Per Pound Sold}}{\textbf{\% Protein}} \times 100$$

= Price Per Pound Protein

The information in this column will reveal some interesting facts about the cost of protein (the most expensive nutrient) especially when you compare different kinds of foods, e.g., dry vs. canned, which have similar values for protein, fat and fiber calculated on a dry weight basis. This column can help you determine which product is really the best protein bargain if you make the calculation using *percent utilizable protein* (amount of protein in a food your dog or cat can really use). To make the calculation for the cost of the usable protein, you will have to write to the manufacturer. Good companies will send you a value for *protein digestibility* and *protein biological value* (the protein's ability to supply your pet with the amino acids he or she needs).

% Digestible Protein × Biological Value × 100 = % Utilizable Protein

(Use this value in your calculation instead of % Protein.)

WHAT TO DO WHEN YOU TAKE A PET FOOD HOME

The final steps in choosing any food for your dog or cat are to take the product home, examine it and feed it. Physical examination of commercial foods gives you very limited information about most products today. Obviously, you should not purchase products in damaged containers or which show evidence of attack by rodents or insects (rodent stool or fur in the food, holes chewed in boxes or packages, insect larvae or dead insects in the package). If all brands of dog or cat food sold in the store are damaged, change your pet food outlet, since the current one is not giving their merchandise proper care. Dry or soft moist foods which show evidence of spoilage (furry white, black, gray or green spots of mold or moist-looking spots of white or yellow bacteria) should never be fed to your cat or dog. Return affected foods to the store and don't feed that particular brand if spoilage is a regular feature.

The ingredients of many canned foods, most dry foods and all semi-moist foods are finely ground so you can tell little about them. However, there are some rules of thumb which can be applied. Avoid products with a darkened surface. In dry foods this may indicate overcooking which destroys valuable vitamins and interferes with protein utilization. Darkened canned foods may also be overcooked or may have reacted with the metal can. (This may or may not interfere with nutrition but is not harmful in itself.) Avoid products containing foreign materials. In canned food this is best determined by spreading some of the food thinly on a plate—large quantities of hair, feathers, wood, bone or other foreign material indicate a poor product with poor quality control. Small pieces of bone are not undesirable. However, large pieces of bone are, as are large quantities of bone, blood vessels, skin or other hard to digest materials. Foods containing these items are, in general, low quality foods which are poorly utilized by a dog or cat and often lead to digestive problems.

The best test of any product you have selected is actually feeding it. No matter how ideal a pet food product looks on paper or how fine its appearance, it can't be nutritious if your pet won't eat it. And if your pet eats it, but develops diarrhea or some other digestive upset when the product is fed, any important nutrients present are likely to be lost.

If your cat or dog develops diarrhea, constipation or becomes flatulent after the gradual introduction of any commercial product, this is not a product you should continue to offer no matter how nutritionally sound it seems or what kind of bargain it may be. New foods introduced suddenly, however, may cause transient digestive upsets. (For more information see page 61.)

What effect a food has on your dog's or cat's stools tells you something about its *digestibility*, and, in a sense, something about its quality and the economy of feeding it. In general, the higher the percentage of the total nutrients absorbed from a food by your pet, the less food is required for proper nutrition and the more economical that food is to feed. A highly digestible food results in stools that are normal in appearance, odor and consistency and that are produced in no more that about twenty-five percent of the volume of the food consumed. Foods with this effect give you the most for your money. Foods which are less digestible (50 to 75% of the volume eaten is passed as stool) still provide satisfactory nutrition for a pet. However, foods which result in voluminous, odorous stools with many food elements visible in them are a waste of your money and feeding them is a disservice to your pet. For more specific information about a pet food's digestibility than you can acquire by watching your pet's body response to eating it, you will have to write to the manufacturer.

VITAMIN-MINERAL SUPPLEMENTS FOR PETS

The most important thing for any pet owner to know about vitamin-mineral supplements for dogs and cats is that they are *not* necessary when a *well-balanced* diet is fed. Manufacturers of good quality complete and balanced diets for pets have put forth much effort and invested millions of dollars in research to produce products which do not need supplementation. The addition of any vitamin and/or mineral supplement to them can upset the delicate balance present and may lead to abnormalities of body structure and function. The worst problems arise when unbalanced supplements such as cod liver oil, vitamin D, vitamin A (see pages 16 and 18) calcium carbonate or bone meal are fed. But even when balanced supplements which provide vitamins and minerals in proper proportions to meet known or estimated daily requirements are fed, problems may arise.

Bony problems in cats related to overnutrition with vitamin A have been established for years. Recent research with the large rapidly growing breeds of dogs indicates that overnutrition may play a significant role in the development of bone disease. Not only can over-supplemented dogs develop obvious bone abnormalities, in some cases a general effect on growth is seen—dogs receiving some supplements are smaller than the relatively "underfed" control dogs! Dogs as breeds do not, in general have more stringent vitamin or mineral requirements than dogs as individuals. Unfortunately many breeders and owners assume any bone or coat abnormality must be due to something lacking in the diet and accordingly over-supplement.

There are times, however, when vitamin or mineral supplementation can be beneficial to a pet. During some illnesses supplementation is required. Your veterinarian will tell you when illness makes vitamin and/or mineral supplementation necessary for your pet. If your pet cannot be taught to eat a balanced diet, supplementation will be necessary. In this situation use the information in this book as a guide to appropriate supplementation and ask your own veterinarian for specific recommendations. If you feel you must add vitamins and/or minerals to an already balanced diet, use a cat or dog vitamin purchased from your veterinarian or a good pet shop and follow directions carefully to supply only the minimum daily requirement for each nutrient.

TREATS FOR PETS

Like most people, most dogs and cats seem to enjoy special tidbits that add variety to their diets although often without much nutritional benefit. And most pet owners seem to enjoy giving treats to their pets from time to time. Only a few things bring forth a smile as quickly as the antics a dog or cat puts on in anticipation of a favorite treat or the obvious pleasure of a pet enjoying an unexpected snack. However, care must be exercised in the selection and feeding of treats for pets as it is in the selection of a day to day diet. Some "goodies" can be dangerous to your animal's health and offering treats to the exclusion of other foods can quickly lead to an obese and sluggish pet. Treat foods which do not provide balanced nutrition should make up five percent or less or your pet's total diet.

TABLESCRAPS

Most dogs and cats will prefer tablescraps to commercial dry or soft-moist foods. However, don't let this fact lead you into feeding your dog or cat an imbalanced diet. If you don't feed large quantities of meat or other goodies routinely, they won't be demanded. Treats should be treats, not a pet's total diet and if you're not very careful, pets offered tablescraps daily are often soon eating little else. If you don't mind owning a pet which is very selective about its food, however, and if you are careful to balance your pet's diet, tablescraps can be given as a pet treat or to improve the palatability of the daily diet.

Avoid offering leftovers or tablescraps which you would not eat yourself. Slightly spoiled foods or undigestible scraps such as bone can harm a pet. Good quality tablescraps such as left over cooked vegetables or their juices, meats, eggs, milk, soups or stocks can be used routinely to improve the flavor of commercial products as long as they are blended thoroughly with a balanced meal and constitute less than twenty-five percent of the meal being fed. Do not feed extremely fatty or spicy foods that may upset your cat's or dog's digestion.

BONES

Offering bones to dogs and cats as treats has both its pros and its cons. They can help satisfy a pet's natural urge to chew and are an excellent natural means of helping to keep a dog's or cat's teeth clean. However, too much bone chewing quickly leads to worn down teeth (a problem primarily for dogs). And bones which are eaten can cause digestive upsets ranging from simple stomach aches to severe constipation or intestinal blockage or perforation.

If you choose to give bones to your pet, be sure they are an appropriate size. They should be large enough and hard enough that they can't be eaten, only chewed on. For a small dog or cat this may mean that a ringbone from a lamb chop is large enough. For larger dogs marrow bones or knuckle bones are the best choices. Parboiled or roasted bones are most healthful in the sense that the heat treatment helps kill any harmful bacteria or parasites that may be present. Do not offer poultry bones, pork chop bones or any other bones which splinter. If you see that you pet is *eating* a bone, take it away. Nine times out of ten an eaten bone will pass along a pet's intestine without harm. However, many pets have died following a gut obstruction or perforation caused by a bone. It's foolish to play the odds when something as valuable as a pet's life is involved.

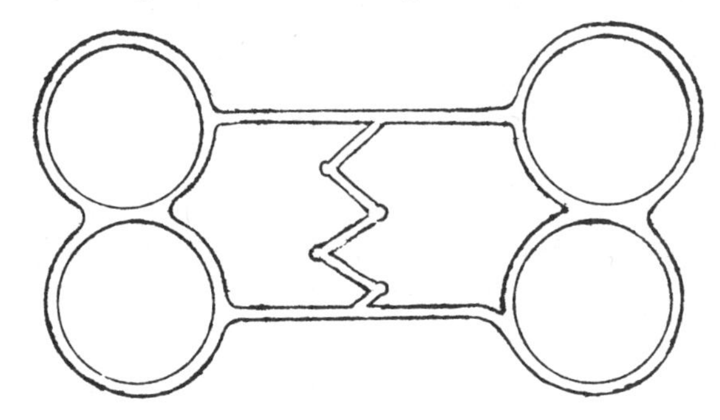

Other items which can satisfy a pet's urge to chew more safely than bones are hard rubber toys, rawhide toys and some vegetables such as carrots. Most cats don't care for these things, but most dogs enjoy them and they are good bone substitutes to offer when a pet's teeth are being rapidly worn down by bone chewing.

HOMEMADE TREATS

Many kinds of mass produced treats are available for pets. A few good ones are designed with the idea of complete and balanced nutrition in mind (look at the labels) and provide the owner with a convenient way to offer his or her pet a nutritious goodie or reward. Most commercial treats, however, are not designed to provide balanced nutrition because they are not meant to compose a major part of a pet's diet. And no commercial treat is a bargain in nutrition; they are not necessities and they are priced accordingly. Artificial colors and preservatives are used extensively in the production of some commercial treats. Many need a long shelf life and all must appeal to your eye in one way or another—who is going to give their dog or cat a treat that they themselves find unappealing?

If you have the time and the inclination to do a little cooking, making homemade treats for your dog or cat can result in some healthful snacks for your pets which are less expensive than the gourmet treats found at your local market or pet store. The following recipes are for home prepared treats that have been pet tested.

This recipe makes an easily stored and transported treat that both you and your pet can eat.

Jerky For Pets and People

1 pound lean beef or lamb. Remove any excess fat. Slice the meat diagonally across the grain into strips 1/8 to 1/4 inch thick.

Marinate the strips for two hours in:
1/4 cup salad oil
1/2 cup soy sauce
1 large clove garlic crushed
1 tablespoon brown sugar

Remove meat from marinade and sprinkle lightly with salt and pepper.

Place meat strips in the oven on a wire rack over a drip pan or hang strips over the oven rack bars with a drip pan placed below.

Heat oven to 175 degrees F

Leave oven door slightly ajar and dry strips for approximately five hours.

Remove jerky when dry but still slightly flexible. Cool and store in an airtight container in a cool place.

Pastries for Cheeselovers

6 tablespoons margarine or five tablespoons lard, bacon fat or vegetable oil

½ cup finely grated cheddar cheese
1 cup all purpose flour
1 small garlic clove minced or mashed

Blend fat and cheese until smooth. Stir in garlic. Then mix in flour. Mixture will be somewhat crumbly.

For Dogs: Shape mixture into a log about 1½ inches diameter and chill until firm.

For Cats: Shape mixture into a log not larger than ½ inch diameter and chill until firm.

To Serve: Cut slices about ¼ to ½ inch thick from rolls. Place on an ungreased cooky sheet. Bake at 375 degrees F until slightly brown (about ten minutes). Makes about two dozen.

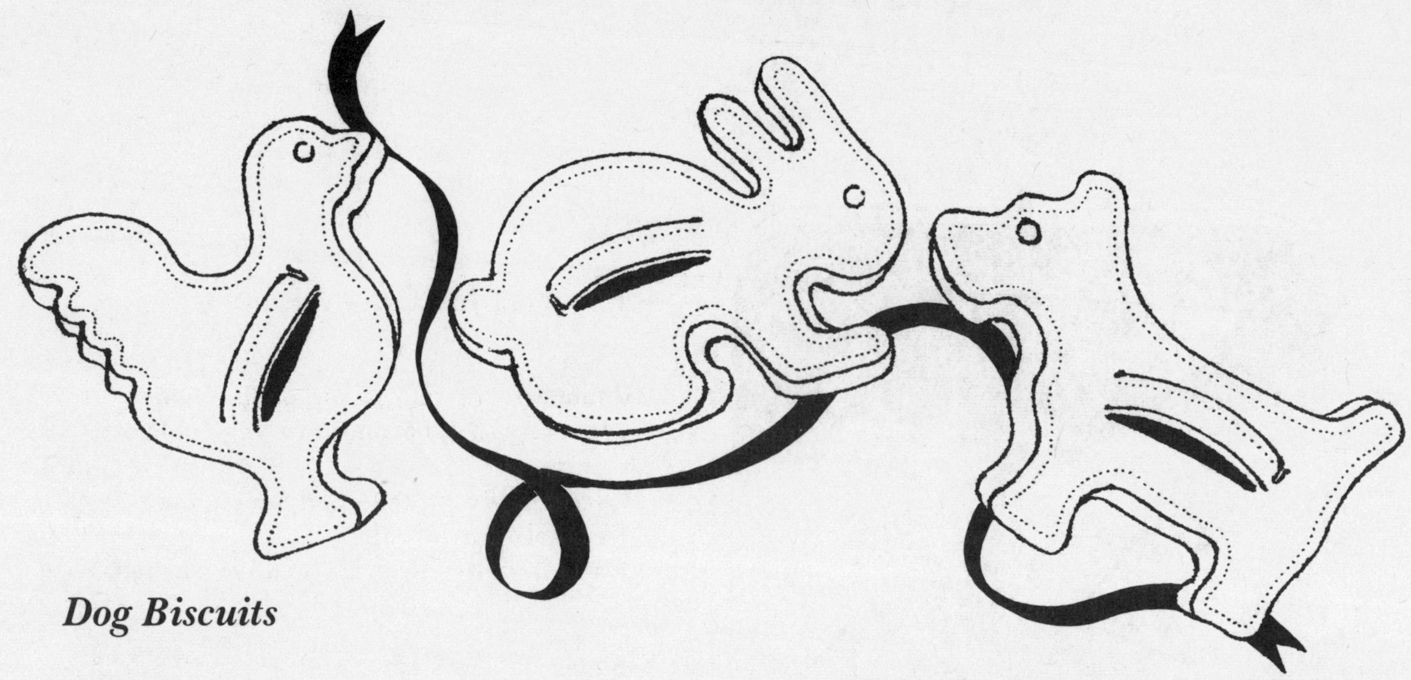

Dog Biscuits

1 cup all purpose flour
1 cup wheat flour
½ cup wheat germ
½ cup powdered dry milk (or soymilk powder)
½ teaspoon salt
6 tablespoons margarine, lard, chilled bacon fat or other shortening
1 egg
1 teaspoon brown sugar

Combine white flour, wheat flour, wheat germ, powdered milk and salt in a bowl.

Cut in shortening until mixture resembles corn meal.

Beat the sugar with the egg. Then stir in the sugar-egg mixture. Add water gradually as necessary to make a stiff dough (approximately ½ cup). Knead on a floured board until dough is smooth and pliable. Then roll out to ½ inch thickness and cut with a cookie cutter (bone or cat shapes are appropriate).

Bake at 325 degrees F until lightly browned (about 30 minutes). Makes about two dozen biscuits.

For flavor variety and color add:

Liver powder (found at health food stores) - 3 tablespoons, or

Dried vegetable flakes - 1 tablespoon, or

Mashed or pureed cooked green vegetables or carrots - 1 cup

Add these flavor ingredients following the sugar-egg mixture and adjust water additions to the dough accordingly.

Cat Munchies

1 cup brown rice
2 cups well-flavored beef, chicken or fish stock (can be made from boullion cubes)
Choose the flavor your cat likes best.

Bring broth to a boil. Stir in rice, bring back to a boil then simmer, stirring occasionally until all broth is absorbed, about 40 minutes.

Spread cooked rice on a cookie sheet.

Bake at 400 degrees F until rice is brown and crackly (about 20 minutes). Or allow to dry at room temperature twenty-four hours.

Heat two to three tablespoons oil in a frying pan. Add rice about ½ cup at a time to hot oil and shake and stir until the grains puff.

Drain on a paper towel. Then serve or store in an airtight container.

Of course there are many other goodies for pets that require minimal preparation. Any meat sauteed and cut into pieces pleases most cats or dogs. A small bowl of cottage cheese, a hard-boiled egg, or a square of cheese are also usually appreciated and are healthful. Avoid offering sweets. Although most cats dislike sweet things, most dogs love them and it becomes easy to end up owning a fat dog with a sweet tooth. Also remember that treats are treats. Unless they are nutritionally complete and balanced, keep the amount fed small.

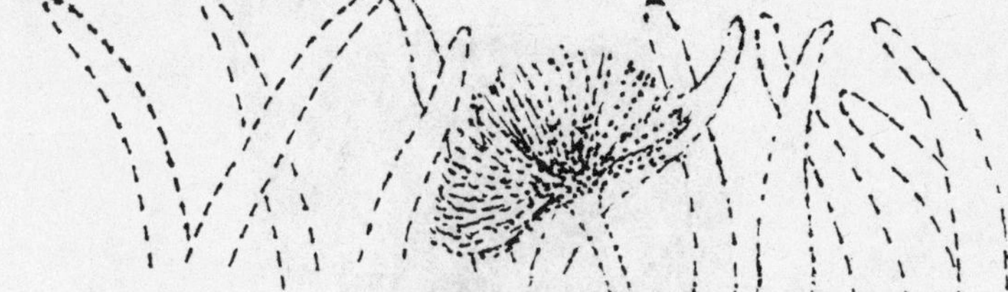

HERBS AND GRASSES

Most cat owners who also keep house plants have found their pets chewing contently at a prized plant's leaves at one time or another. While most dogs seem to leave treasured house plants alone, we have all seen dogs eating grass outdoors. This plant eating behavior is often attributed to an upset stomach, and vomiting does, indeed, frequently occur when a pet eats fresh greens. Although many dogs and cats will seek greens when their stomachs are upset, vomiting following plant eating is often due to the irritating characteristics of the plants themselves rather than the presence of any illness. Coarse bladed grasses and slightly poisonous houseplants can easily upset the healthiest stomach.

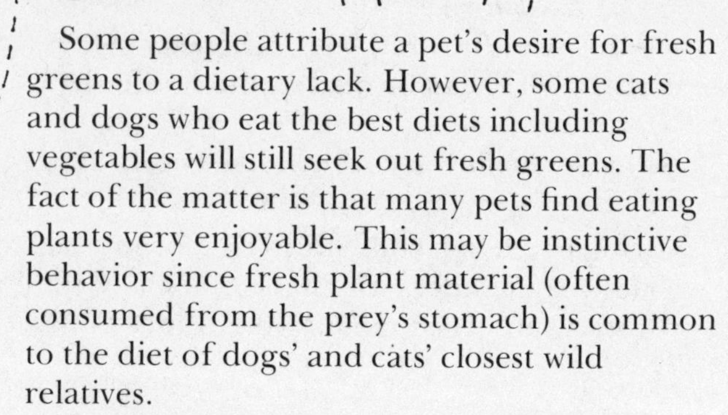

Some people attribute a pet's desire for fresh greens to a dietary lack. However, some cats and dogs who eat the best diets including vegetables will still seek out fresh greens. The fact of the matter is that many pets find eating plants very enjoyable. This may be instinctive behavior since fresh plant material (often consumed from the prey's stomach) is common to the diet of dogs' and cats' closest wild relatives.

Whatever the reason cats or dogs have for eating growing plants, there is no question that you can make many pets happy by providing a patch of greens for them to nibble. Order a bunch of ryegrass seeds (wheat is also good) from your nursery or seed catalog and sow a small green plot in your garden for your pet. If you are an apartment dweller or your pet stays indoors, grow a patch in a container placed on a sunny windowsill. (You can buy preplanted containers at pet stores or specialty shops.) Let your pet munch on the greens while they are young and tender or harvest the tender shoots with scissors and put them in the feeding dish.

CATNIP

A special green treat for cats is the herb *Nepeta catalaria*, commonly called catnip or cat mint. This herb is a member of the mint family and like most mints is easy to grow outdoors (or as a houseplant if given the proper conditions). In fact, catnip grows as a weed throughout most of the northern Midwest and lower Canada. Catnip plants are sold at many nurseries and catnip seeds are available through seed companies, nurseries and in kits sold at pet stores and specialty shops. For cat owners who don't garden but still want to offer their pets the joys of catnip, catnip stuffed toys, dried catnip and even an aerosol containing the active volatile oil in catnip (nepetalactone) are sold. Not all cats seem to enjoy catnip (somewhere between 30 to 50% of all cats ignore it), but those who do seem to derive real pleasure from its effects.

The catnip reaction starts when a cat sniffs the herb. This is quickly followed by licking or chewing the plant, (especially if it fresh) and head shaking, salivation and rubbing the head against the catnip source. Some cats gaze into space. Cats which respond most intensely to catnip will rub their whole bodies over it and roll and stretch along the ground. Catnip toys often elicit mock pre-killing behavior and leaping about. Frequently the catnip reaction mimics some aspects of courtship and breeding behaviour, especially the rolling and rubbing of a female in heat. Some cats even seem to have hallucinations and spend time chasing phantom butterflies or mice. After roughly three to fifteen minutes the catnip reaction is over.

Exactly how catnip induces such signs of pleasure in some cats and why others don't respond to it is not known. However, the active agent in catnip is structurally related to both LSD and the active agent in marijuana so reactor cats may experience some kind of psychedelic high. Although dogs are known to be attracted to the smell and taste of marijuana and its derivatives, catnip has no apparent lure for them.

CHAT
DR. TERRI McGINNIS'
GOOD FOOD
SALT
Baby

WHAT & HOW TO FEED YOUR PET

Cats and dogs alike have special requirements for feeding and nutrition during each life stage. Not only must you select the proper foods to meet any specialized nutrient needs your pet may have while young, as an adult or during old age, but the way you feed a pet can be as important as what foods you offer.

Healthy cats and dogs are not naturally finicky eaters. Although scientific studies have shown that cats tend to be more selective about their foods than dogs and that some dogs are more selective than others, these pets, contrary to what some advertisements might lead us to believe, do not naturally have to be coaxed to eat. Dogs and cats base their food selections on a combination of texture, odor, taste and the effect of early experiences with food. There is no question that some foods are naturally more palatable than others to pets, but as you have already learned, the foods most tasty to cats or dogs are not always the most nutritious for them. And cats and dogs given free choice of foods will not automatically select a diet that fulfills their nutritional requirements. Any inborn food preferences of dogs and cats are usually easily shaped by early food experiences. Like human gourmets, however, pets accustomed to highly palatable gourmet diets have difficulty adjusting to new, less costly ones. In fact, pets probably have more difficulties adjusting to new foods than people do since they are creatures of habit and develop narrow food preferences easily.

Use the role of early food experiences to shape your pet's feeding behavior towards a nutritious diet. A single commercial food which is designed to be a complete and balanced diet may be fed continually since dogs and cats do not require dietary variety to be happy. However, the best approach is to feed a varied diet while your cat or dog is still young. This kind of feeding insures diverse nutrient sources and helps guard against small vitamin and/or mineral deficiencies. It also encourages the development of very mixed gut flora which helps prevent digestive upsets when new foods are eaten. Pets fed only one kind of food are notorious for developing digestive upsets when new foods are suddenly offered.

HOW TO CHANGE THE DIET

Feeding a varied diet to your pet does not mean that abrupt changes in food are wise or possible. The digestive systems of cats and dogs used to eating an unvarying diet become adapted to that diet and quick changes can often lead to diarrhea and/or vomiting. In the case of pets who have been eating the same diet for weeks or months, changing the diet quickly is often not possible for behavioral reasons.

Approach changing a dog's or cat's diet in a logical manner. First analyze what the diet has been in the past. If the diet has been mainly one class of food—dry, semi-moist or canned—a quick change to a new food in the same catagory can usually be made. Cats or dogs eating dry foods can ordinarily be switched over to semi-moist foods and vice versa. Both kinds of food are relatively high in carbohydrate so that gut flora has no major adaptations to make. Problems often arise, however, when dogs or cats eating dry foods are abruptly changed to canned foods (usually canned meats with little added cereal). Since canned foods are highly palatable to most pets they are usually eaten readily, often in large quantities. The intestinal flora is not prepared for such an abrupt and major change in foodstuffs and indigestion usually follows. In theory changing from canned to dry foods can cause similar digestive upsets. However, since dry foods are rarely more attractive to pets accustomed to eating canned foods, overeating them does not usually occur when they are offered as a new diet and the digestive system has time to make a gradual adjustment.

Use this method to change an adult dog or cat's diet quickly:

1. Feed approximately fifty percent of the usual diet for one day before the change. This should insure a good appetite the day the new food is introduced.

2. On the day of the change and for one day after it, feed the new food in fifty percent of the quantity recommended.

3. If your pet won't eat any or eats only a small amount of the new food and is in good health, continue to offer the new food for four days before offering food your pet finds more palatable.

Always use a gradual method for changing the diet of puppies or kittens. Their immature digestive systems must be treated with kindness and because they grow rapidly they cannot be allowed to refuse food for days at a time. A gradual method for changing the diet is also necessary when switching from dry foods to canned foods to avoid causing digestive upsets. Use this method for dogs and cats who are really finicky eaters too. A gradual change masks the texture, flavor and odor of a new food with that of the old one until the dog or cat becomes adjusted to it.

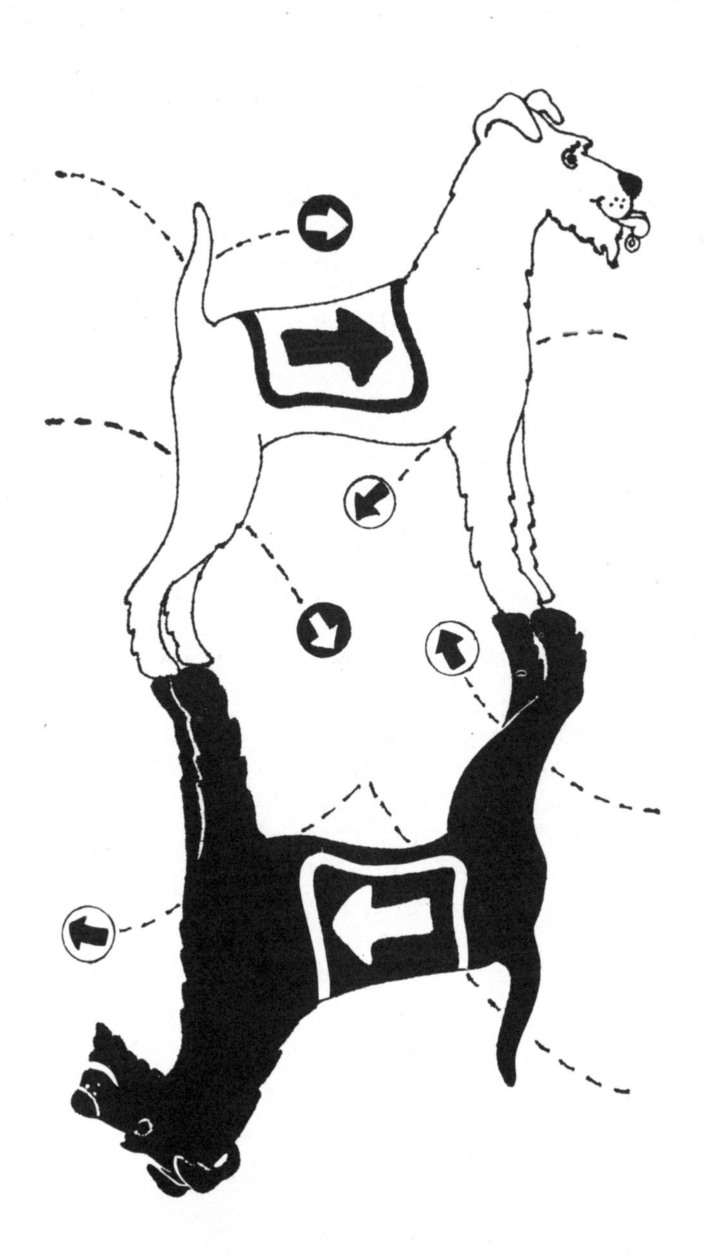

The standard way to change a pet's diet slowly is to replace 25% of the existing diet at a time with the new food. Mix the old and new foods thoroughly making every effort to mask the presence of the new food. When your pet has been eating the initial mixture well for some time, change the mixture to 50:50 old diet and new. The next mixture to offer is 75% new food and 25% old food. Then finally offer only the new diet. The keys to success with this kind of diet change are patience and persistence. You cannot move from one step to the next until your cat or dog is eating the mixture well. This usually means at least two to three days of feeding one mixture before moving on to the next. If a 25% addition of new food seems to be too much, make the changes more gradually.

The most difficult diet change for most pet owners to make is from feeding canned meat type foods to dry foods. The change is facilitated by moistening the dry food while mixing it with the canned one until the change over to eating moistened dry food is complete. Then gradually decrease the amount of liquid mixed in until only the dry product is being fed. Water, stock or boullion, milk, or a gruel made by mixing water with a canned meat product can all be used to moisten a dry food. Warm liquids are better to use than cold ones; they improve the texture and aroma of many foods for pets.

HOW TO FEED A KITTEN OR PUPPY

In order to meet the nutritional requirements of a kitten or puppy for proper growth and development, you must not only provide a diet that supplies 50 to 150% more calories per pound body weight each day (amount dependent on age and breed) than that necessary for an adult, but also approximately twice the protein. This must be done while meeting all the other nutrient requirements a young animal has, requirements certainly at least as strict as those of an adult dog or cat. Frequent feedings will allow a puppy or kitten to meet the calorie requirement if the diet provided is well-balanced and of the proper caloric density. Protein requirements are most easily fulfilled by selecting good quality complete commercial puppy or kitten chows as the basic diet. Other non-special purpose complete foods may also be used if they contain at least 34% or more protein for kittens, 28% or more protein for puppies calculated on a dry matter basis. Eggs, milk and milk products such as cottage cheese, yogurt, and sour cream are high quality proteins which are valuable as dietary supplements for growing animals. However, be sure to avoid any milk product which causes diarrhea when fed (see page 10). Small amounts of cooked muscle meats, fish and liver are also good protein supplements which may be introduced gradually at a young age so your pet will become accustomed to eating a variety of foods. However, unless you are using any food as a supplement to complete a diet not adequate to meet the needs of a growing dog or cat, it should not comprise more than twenty-five percent of the diet or you may cause dietary imbalances that can interfere with the proper growth of your pet.

Be sure to find out what your new pet has been eating before you bring it home and try to offer the same or a very similar diet the first day or two. Although it is very important to get a young animal to eat a well-balanced diet as soon as possible, too many changes at once can cause digestive upsets. Give your kitten or puppy a quiet place of its own where a bed and food and water dishes are all nearby then let it adjust to the new home before making any major diet changes. When you do begin to change the diet, gradually introduce the new foods. Start with a

single complete commercial food and add it to the original diet increasing the quantities of unfamiliar food gradually and decreasing the initial food until the newcomer is eating the proper diet well. Then introduce other new foods in small portions at first to avoid digestive upsets.

Kitten and puppy milk replacers such as KMR®, Esbilac® and Orphalac® can be very helpful when feeding a very young pet. Mixed with commercial foods, they make them easier to chew. They mimic mother's milk and are therefore usually readily accepted. They do not have the tendency to produce diarrhea as cow's milk does, and they provide complete nourishment in themselves so they can be mixed safely in any proportion with other complete foods. Be sure, however, to provide your kitten or puppy with fresh water at all times whether or not liquids such as milk replacers or cow's milk are provided.

It is physically impossible for a puppy or kitten to consume enough food (even of the highest quality) at one sitting to meet the daily calorie requirement. The simplest and most convenient way to be sure that your pet is consuming enough to meet his or her daily caloric needs is to allow *self-feeding* of a balanced complete commercial diet. To use this method leave the food out where the youngster has free access to it and change it as necessary to keep it fresh. Unless the food is too highly palatable, most kittens or puppies will not overeat with this feeding method. In most cases the amount dogs and cats choose to eat is only that necessary to fulfil their daily calorie needs. Not only is self-feeding convenient for the pet owner but it also seems to help prevent boredom in pets who must be left alone during the day. Self-feeding must be abandoned or the portions left out reduced if your pet tends to become too fat.

Scheduled or *portion controlled* feeding is the system where you provide your kitten or puppy with several meals daily and do not offer food between meals. It usually results in a pet that is anxious and ready to eat at mealtimes making it easy for you to determine when the appetite is not normal. It can, however, produce a dog or cat that is "too attuned" to food with a tendency to gorge at mealtimes and to become fat. If you choose a scheduled feeding method provide four meals a day until your pet is three months old, three meals a day until six months of age, then two meals a day until adulthood.

A combination of self and scheduled feeding may be used instead of relying on one system or another. Many people successfully leave out a variety of commercial complete dry foods for their pet's free choice consumption and use the more palatable complete canned products alone or mixed with other complete foods as "treats" or scheduled meals. Unbalanced supplementary foods such as eggs and meat should always be mixed with or supplemented to make a complete food when fed as meals for young animals. Dogs and cats who become used to having extra palatable foods at scheduled meals may fail to consume the balanced diet left out for their free choice feeding.

HOW MUCH TO FEED A KITTEN OR PUPPY

You can use the calorie tables on page 79 as rough guides to estimate your kitten's or puppy's daily energy needs. If you don't know the calorie content of the food you feed, write to the manufacturer for the information or use the feeding guides on the food package to help you determine your pet's daily needs. Remember, however, that each animal is an individual and, as such, has individualized calorie requirements. Body size (breed), activity, environmental temperature and other factors all affect a kitten's or puppy's daily food needs. Your pet's appearance (while young or as an adult) is one measure you must always consider no matter what other guides you use to judge the adequacy of the diet.

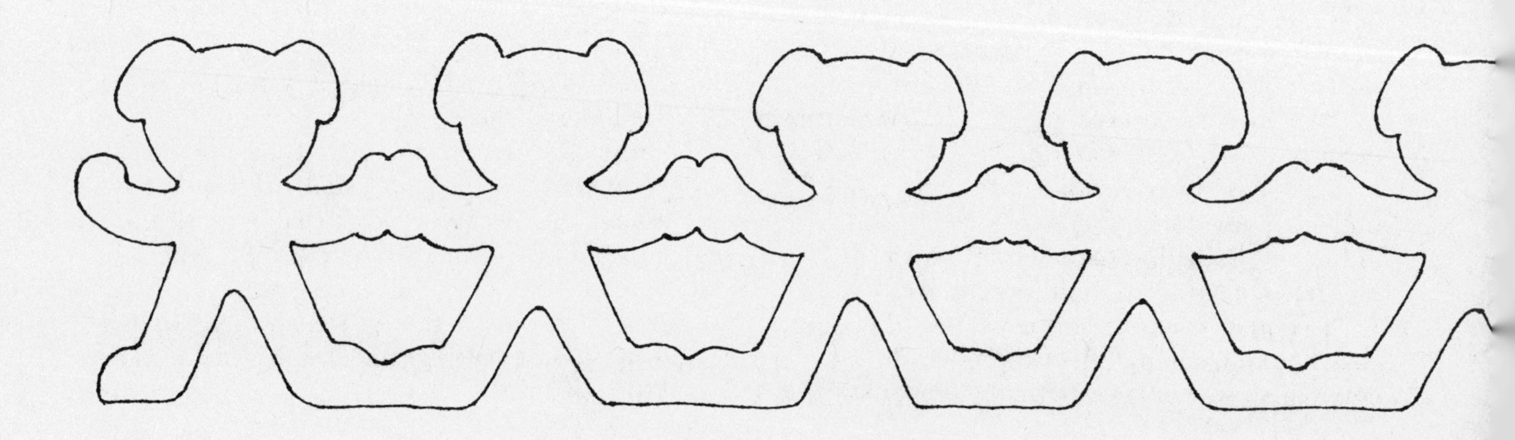

Look at and feel your pet. If the spine and ribs are prominent, you may not be feeding enough. A glossy coat, free from dandruff, good health and activity are all signs that tend to indicate that enough of a nutritious diet is being fed. Weigh your kitten or puppy and plot its growth curve. A steady weight gain is to be expected if your pet is healthy and well fed. Be sure to make a distinction, however, between proper growth and excessive growth. Recent nutrition studies imply that obesity is undesirable in young animals as it is in adults and that bone abnormalities in some dogs may be related to overfeeding and too rapid growth. If your dog's growth curve is too steep as compared to the standard for its breed or if your cat or dog is chubby, discuss a diet change with your veterinarian. Poor growth, poor coat or frequent illness also call for a veterinarian's examination and could mean that your pet's diet is inadequate.

If you are using scheduled feeding, each meal should comfortably fill a puppy or kitten. If the stomach is very distended and taut following a meal, or if vomiting occurs shortly after eating, you may be feeding too much at one time. More frequent smaller meals may be necessary. If your pet is still willing to eat after a meal and tends toward thinness, you may need to offer a larger portion at each feeding. Unless your cat or dog has a tendency to overeat, scheduled meals should always be offered in large enough portions that some food is left in the bowl.

Any dietary problems with feeding young animals not quickly resolved at home (within twenty-four to thirty-six hours) should be discussed with a veterinarian. Because of their rapid growth, small size and relatively high metabolic rate, what sometimes appear to be minor dietary problems in puppies or kittens can cause severe illness to develop quickly.

On the following pages you will find charts which depict the average growth by weight of cats and various dog breeds.

GROWTH CHART FOR DOGS:

The curves on this chart represent the growth of normal, healthy individuals of fifteen dog breeds. Note that all dogs fall roughly into four groups based on their eventual adult size (toy, small to medium, large and giant) and the age at which they reach their mature body weight (point where the curve levels out). Since only a rare individual's growth will exactly match that of a standard curve and since many pets are mixed breeds, use these curves only to give yourself a rough estimate of your dog's expected growth rate.

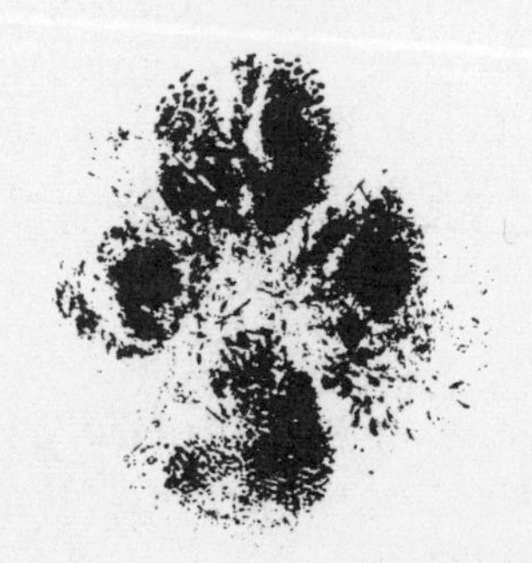

Modified and reproduced with permission given by the editor and publisher from Current Veterinary Therapy, R.W. Kirk, ed., copyright W.B. Saunders Company 1974.

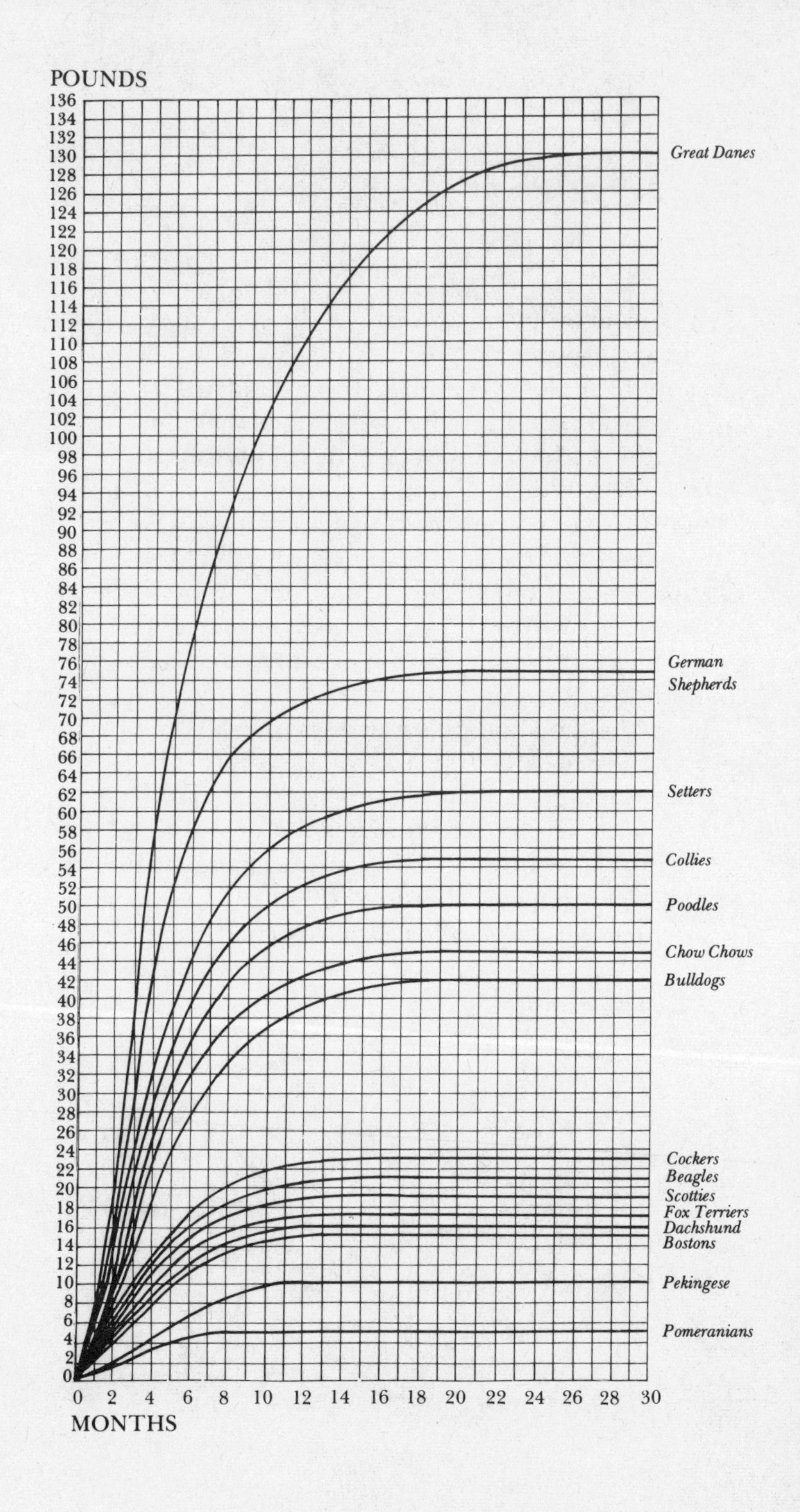

GROWTH CHART FOR CATS:

This chart illustrates the average growth of kittens during their first year. The curve is based on equal numbers of male and female cats; expect a male kitten to grow somewhat more rapidly than a female. Since no individual's growth can be expected to exactly match an average curve, use this chart only as a guide to your pet's expected growth.

BIRTH TO ONE YEAR

BODY WEIGHT

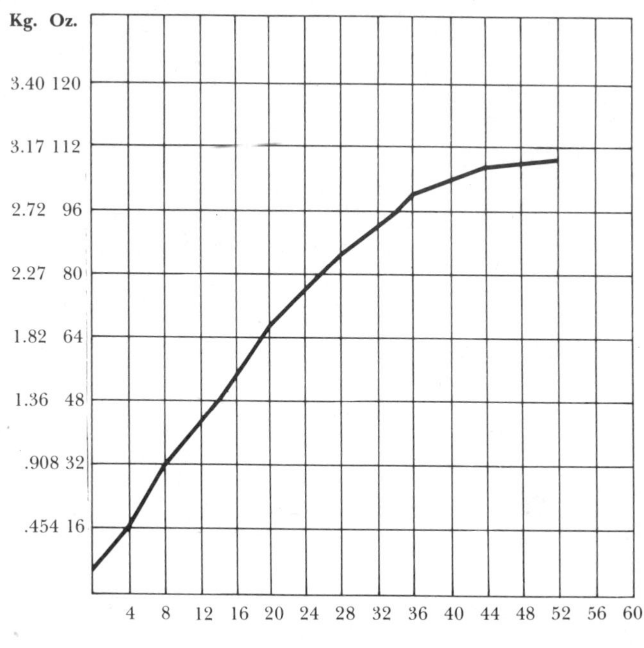

AGE IN WEEKS

Compiled from growth charts used with permission of the Ralston Purina Company, copyright 1975.

PLOT YOUR PET'S GROWTH:

Between the ages of two to six months the growth of all kittens and puppies is relatively rapid. Since small dogs and cats tend to grow to be small adults and large young ones tend to become large adults, direct comparison of your pet's growth to the standard growth charts can

PUPPY GROWTH CURVE*

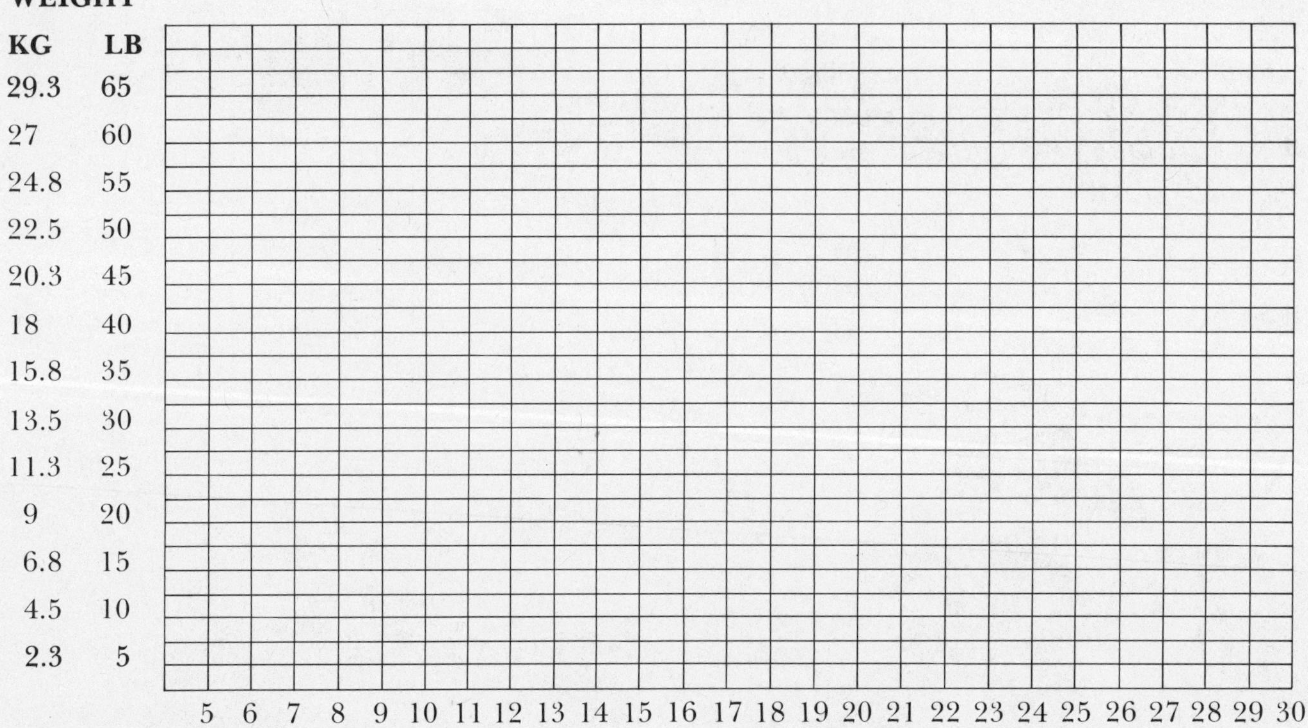

**If your puppy will weigh ten pounds or less as an adult, plot his or her growth on the kitten growth chart.*

be misleading if your dog or cat tends naturally to be smaller or larger than average. It could also lead to harmful overfeeding if you attempt to force a small pet's weight up to meet a standard curve. Plot your kitten's or puppy's growth weekly on the charts provided below. A relatively straight growth line between the ages of two and six months (three to five months for dogs weighing more than sixty pounds as adults) is normal. For dogs, the slope of the curve (steepness should be similar to that standard for your pet's breed or a breed which reaches a similar adult size. A curve that is more or less steep than average may indicate over or under feeding. (See page 58 for standard growth curves for fifteen dog breeds.)

KITTEN GROWTH CURVE*

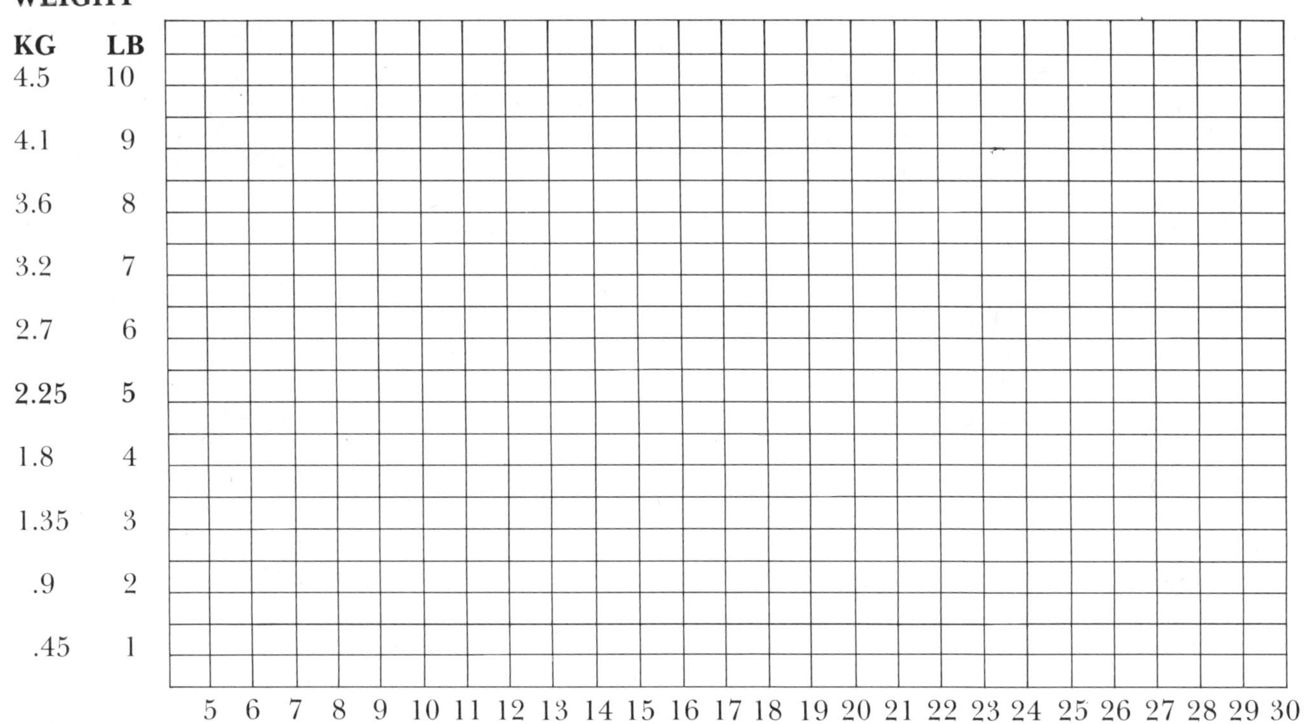

Use this chart to plot your puppy's growth if he or she will weigh ten pounds or less as an adult.

HOW TO FEED AN ADULT CAT OR DOG

When your pet's growth curve has leveled out and he or she is mature you can switch from a puppy or kitten ration to an adult food. Self-feeding or combination feeding can be continued, but if you notice your pet becoming overweight, you will have to switch to one or two individual meals per day. Cats seem to prefer two meals, however one meal is adequate for most cats and also for most dogs. Some pets, dogs in particular, vomit clear or yellow-tinged stomach juices if fed only once every twenty-four hours. These animals will benefit by being fed two smaller meals (breakfast and dinner) each day.

Most adult dogs require about forty calories per pound body weight each day. Most cats require about fifty calories per pound body weight each day. However, as with young animals, each individual has individualized calorie requirements and any feeding information (given here or on a food package) can only be used as a guide. Active cats may require 50 to 100 percent more calories each day than sedentary ones. Uncastrated males require more calories than neutered one. During hot or cold weather dogs kenneled outside may need from 20% to 100% more calories each day to maintain a proper body weight. Pregnancy and lactation (nursing) impose their own special requirements on female cats and dogs (see page 82).

In most cases any good complete and balanced commercial food will meet the daily calorie and nutrient requirements of a house pet. Complete and balanced rations can be mixed together in any proportions to meet your pet's needs. Remember, however, that high quality unbalanced protein foods used to supplement commercial products should comprise less than twenty-five percent of the diet. Snacks and tidbits that do not have good nutritional value should compose less than five percent of the diet. Specialty foods or a switch from one type of commercial food (e.g., dry food) to another type of balanced commercial food (e.g. canned meat ration) or from feeding a single type to a mixture of foods may be necessary when a house pet's life changes to require more energy. An example of this situation is found in the case of a dog which is used for hunting part of the year and is relatively sedentary the rest of the year. During hunting season a dog like this may be able to meet the increased calorie needs merely by eating more of the regular dog food. However, when this usual food does not have enough calories per volume consumed (too *low* an *energy density*), the hunting dog cannot eat enough to meet the energy requirement and/or gets a digestive upset (usually diarrhea) while trying. A switch to a food which contains more calories per pound fed (food with a *higher* energy or calorie density) usually solves the problem. An analogous situation is that of the tomcat who eats a highly palatable, energy dense canned food all winter and is roly-poly when spring comes around. If his diet is not changed to include more food, a remarkable weight loss ensues as the tom uses more energy to fight and mate during the spring breeding season. Many cat owners have thought their tom was sick or "wormy" when this unexpected weight loss was seen.

FEEDING THE OLDER CAT OR DOG

Dogs and cats undergo aging changes just as humans do and may require special diets for maximum health and activity in old age. In general, older animals require fewer calories per pound body weight each day as compared to their needs while young; the amount of food given must usually be decreased in order to avoid causing obesity. Some body changes can result in decreased utilization of nutrients; additionally, intestinal absorption of nutrients may be impaired. There is then a rationale for the occasional recommendation to use a balanced vitamin-mineral preparation to supplement an older cat's or dog's diet. For older dogs who seem to have trouble digesting fats or who tend to become obese easily, dietary fat should be kept to the minimum. (Choose foods with no more than six to eight percent fat, dry basis.) However, most cats and dogs live healthfully to old age eating their usual adult diet fed in somewhat decreased amounts.

Certain diseases such as heart, kidney or liver failure, which tend to occur more often in older animals, require special diets. Only your veterinarian can diagnose the presence of such conditions. If such a health problem is diagnosed in your older cat or dog, your veterinarian will make suggestions for appropriate dietary changes.

On the following pages you will find sample diets which will provide good nutrition for cats or dogs.

A NUTRITIOUS DIET FOR DOGS

FEED DAILY:

Complete and balanced commercial dry dog food. Products containing less than 8% fat need fats added (corn or safflower oil are good). Feed once or twice a day or allow free access.

and/or

Complete and balanced canned food. Canned foods are best fed mixed with dry foods or fed only once or twice a week. Complete canned foods are nutritious; however, their palatability may encourage overeating and their composition seems to encourage tartar formation on the teeth.

Water should be available at all times.

FEED OCCASIONALLY:

Eggs, milk, cheese, yogurt or other dairy products, nutritionally incomplete canned meats, cooked vegetables, cereals, cooked fish. These extras should not constitute more than 25% of the diet. Beef liver can be offered once a week.

Dogs may also be offered brewer's yeast and other "people foods' such as fruits, uncooked vegetables, sweets and condiments. Remember, however, that the last listed items may cause digestive upsets and they do not contribute significantly to nutrition. If offered, they should constitute less than 5% of the diet.

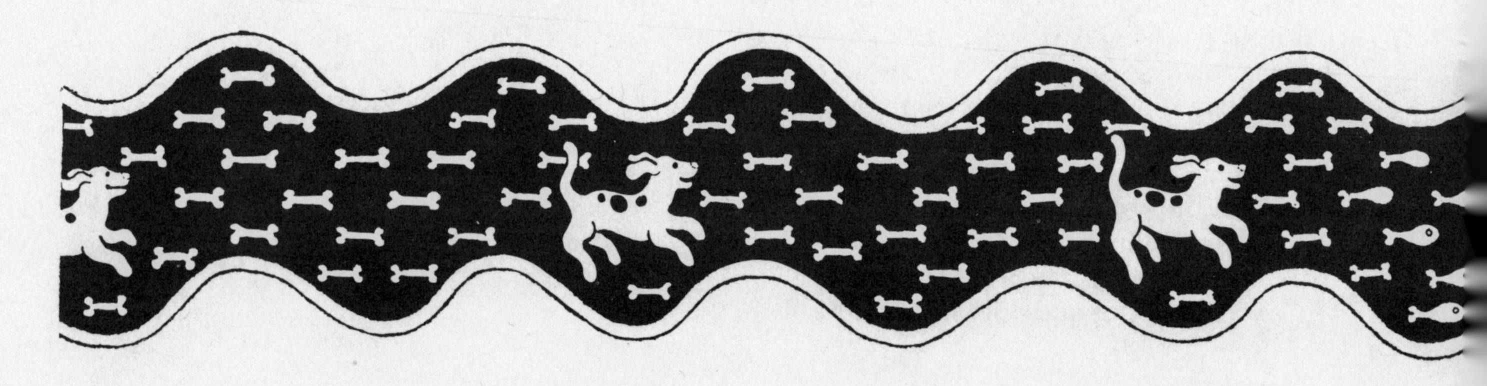

A NUTRITIOUS DIET FOR CATS

FEED DAILY:

Complete and balanced commercial dry cat food. Products containing less than 10% fat need fats added (corn or safflower oil are good). Feed once daily or allow free access.

and/or

Complete and balanced canned food—offer about one can per seven pounds body weight if other food is not given.

Water should be available at all times.

FEED TWICE A WEEK:

Beef liver—once ounce per adult cat. Organ meats (spleen, kidney) may be substituted for liver but fail to provide the high level of vitamins and minerals that liver does. Lightly cooking meat products helps prevent parasite transmission without destroying important vitamins.

FEED OCCASIONALLY:

Cheese, yogurt, sour cream, milk, cooked vegetables, eggs, soups, cooked cereals, baby foods, brewer's yeast, cooked clams or fish.

Cats may also eat other "people foods" such as fruits, uncooked vegetables, sweets and condiments as treats if they do not cause digestive upsets; just remember that such foods do not contribute significantly to a cat's nutrition.

CATNIP
Angel
FEBRUARY
Kitty
OF MICE & CATS
OPEN
MOO
VITAMIN D
CATSUP

SPECIAL NEEDS–SPECIAL FEEDS

THE FAT CAT OR DOG

The overweight cat or dog almost always becomes so as a result of the owner's good intentions. It's difficult for most pet owners to resist sharing the special treats they enjoy with their cat or dog, especially when he or she sits nearby looking longingly at the goodies. It is also difficult to resist feeding a pet too much of his or her regular diet when it is highly palatable and the appropriate portion disappears quickly at mealtimes. Unfortunately, unless some real health problem exists, obesity occurs only when more calories are consumed than are used in the day's activities. And the only way to reduce the weight of a fat cat or dog is to reverse this situation.

Some roly-poly cats and dogs are quite engaging. However, any pet (young or adult) which is so heavy that you cannot feel the ribs under a relatively thin layer of skin and fat is overweight. Their cute external appearance can conceal serious internal problems caused by the extra pounds they carry.

Excessive body fat puts excessive strains on the joints, heart and lungs. A veterinarian has more difficulty examining and treating an obese animal than a normally-fleshed one since excess fat interferes with listening to or feeling the heart beat and with feeling the pulse and the abdominal organs. Obese pets are poorer surgical risks and it has been scientifically shown that fat pets (dogs) are more susceptible to viral and bacterial infections. Most overweight dogs and cats become inactive, changing from the young, sprightly companion you once knew to the sluggish pet that sleeps away the day on a pillow.

The best way to approach the problem of obesity in a pet is to prevent it by monitoring your dog's or cat's diet carefully from the time you first begin feeding. However, if your pet is already too fat, don't despair. A few calculations followed by changes in the diet can remedy the situation. A veterinarian's examination can determine whether your dog or cat is in good general health and whether or not the excess weight is caused by an hormonal imbalance (this requires blood tests). If your pet is healthy except for the excess weight, put him or her on a diet.

Don't try to rely on increasing your pet's exercise to achieve any weight reduction. It takes lots of exercise to result in a single pound of weight loss and lack of exercise is rarely the cause of fatness in pets. Do plan to change your pet's diet permanently; only a new feeding plan can prevent pounds lost from returning. This is why reducing diets for pets can be disappointing. If you feed your pet a reducing diet only until the desired weight is reached, then go back to the old feeding pattern the lost weight will return. So use special reducing diets only if you fail to achieve weight reduction by feeding appropriate quantities of a normal, nutritious diet and only if you can live with the possibility of having to feed a special food indefinitely to maintain your pet in good health.

The best reducing diet for a cat or dog is a complete and balanced one which is high in protein, contains all the necessary vitamins and minerals and contains just enough fat to meet the daily minimum requirement for health. This type of diet then can be continued as the usual food. No fattening tidbits should be offered, but if your dog or cat is accustomed to begging and you can't resist, offer low calorie treats such as pieces of raw vegetables (carrots are good for dogs) or a clean bone to chew on. Dividing the allowed quantity of food into many small portions is also helpful.

Choose the weight you want your pet to reduce to then consult the appropriate calorie chart to determine the approximate number of calories your dog or cat needs each day. Then feed *fifty* to *sixty* percent of the daily calorie requirement until the desired weight is reached. This could take several weeks.

APPROXIMATE DAILY CALORIE REQUIREMENTS FOR DOGS:

Desired Weight		*Calories Needed For Maintenance*
Lbs.	*Kg.*	
5	2.3	250
10	4.5	420
15	6.8	525
20	9.1	700
30	13.6	960
40	18.2	1200
50	22.7	1350
60	27.3	1500
70	31.5	1750
80	36	1960
90	40.5	2160
100	45	2300

APPROXIMATE DAILY CALORIE REQUIREMENTS FOR CATS

Desired Weight		*Calories Needed For Maintenance*
Lbs.	*Kg.*	
4	1.8	120
5	2.3	150
6	2.7	180
7	3.2	210
8	3.6	240
9	4.1	270
10	4.5	300
11	5	330
12	5.4	360
13	5.9	390
14	6.3	420
15	6.8	450

DAILY CALORIE REQUIREMENT

AGE:	DOG*		CAT	
	Approximate Calories Needed Per Pound	*Approximate Calories Needed Per Kilogram*	*Approximate Calories Needed Per Pound*	*Approximate Calories Needed Per Kilogram*
0-1 week	60	135	190	415
1-2 weeks	70	155	125	275
2-3 weeks	80-90	175-200	120	265
3-6 weeks	90-120	200-265	110	220
2-3 months	100-90	220-200	95-70	210-155
3-5 months	90-65	220-145	70-50	155-110
5-7 months	65-50	145-110	50-45	110-100
7 months-Adult (dog or neutered cat)**	50-40	110-90	40 or less	90 or less
Active adult dog or tom-cat	Two times maintenance needs	Two times maintenance needs	about 50	about 110
Pregnant female	1-1.3 times maintenance needs	1-1.3 times maintenance needs	about 50	about 110
Lactating female	1.5-3 times maintenance	1.5-3 times maintenance	125	275

**Dogs' calorie needs tend to vary inversely with their body weight. Small dogs need more calories per pound each day than large ones do so the values here can only be used as a rough guide. Calculate dogs' needs using the higher values until you can judge by other criteria what your pet's daily needs are.*

***For more specific guidelines to adult calorie needs see the calorie charts on the preceding page.*

Use the following chart as a guide to calculate how much standard commercial food will provide the proper number of calories for your pet's reducing diet or write the manufacturer for more exact calorie information if it is not printed on your pet food's label.*

Type of Complete Food	*Approx. Calories Per Lb. Dog Food*	*Approx. Calories Per ¼ lb (4 oz) Cat Food*
Dry Food	1500	400
Semi-moist	1350	400
Canned meat	600	125
Canned mixed *(meat and cereal or soy products)*	500	90

Remember, once you have done the calculations and the new diet is begun *more food is not allowed*. If you find yourself giving in to the urge to treat your pet be sure to include calorie values for any tidbits you've given.

Weigh your pet weekly. If you are following the rules herein and your dog or cat is not losing weight, consult your veterinarian for further help. Some individuals may need fewer calories than the approximate values suggested. Once your pet has reached the desired weight you can relax the rules a little to increase the calorie level to the maintenance amount for that weight.

**If you make your pet's food you will have to determine the calorie content yourself.*

AN EXAMPLE: Your cat weighs twelve pounds (5.4 kg), but should weigh nine (4.1 kg). The daily maintenance calorie requirement for a nine pound cat is 270 calories × 60% = 162 calories to be fed while reducing. This is about 1.6 ounces dry or semi-moist food, 5 ounces canned meat, or 7.2 ounces of a canned mixed diet. After the desired weight is reached, feeding for maintenance would be about 2.7 ounces dry or semi-moist food, 8 ounces canned meat or 12 ounces canned mixed diet.

AN EXAMPLE: Your dog weighs thirty pounds (13.6 kg), but should weigh twenty (9.1 kg). The daily maintenance calorie requirement for a twenty pound dog is about 700 calories × 60% = 420 calories to be fed while reducing. This is about 13 ounces of canned mixed complete diet, 11 ounces meat, 5 ounces semi-moist food or 4.5 ounces of dry food. When the desired weight of twenty pounds is reached, the food intake could be raised to about 20 ounces canned mixed diet, 18.5 ounces canned meat diet, 8.4 ounces semi-moist food or 7.5 ounces dry food.

A ten pound cat or dog needs less than 450 calories energy each day. The addition of any of the following treats to a diet appropriate for this size pet will cause the food intake to exceed the daily calorie need by 25% to 50%: 1 tablespoon butter, 1 cookie, ½ donut, 10 potato chips, 2 slices bacon, ½ hot dog, ½ cup ice cream. Clearly, you must be careful about giving treats to a dieting pet.

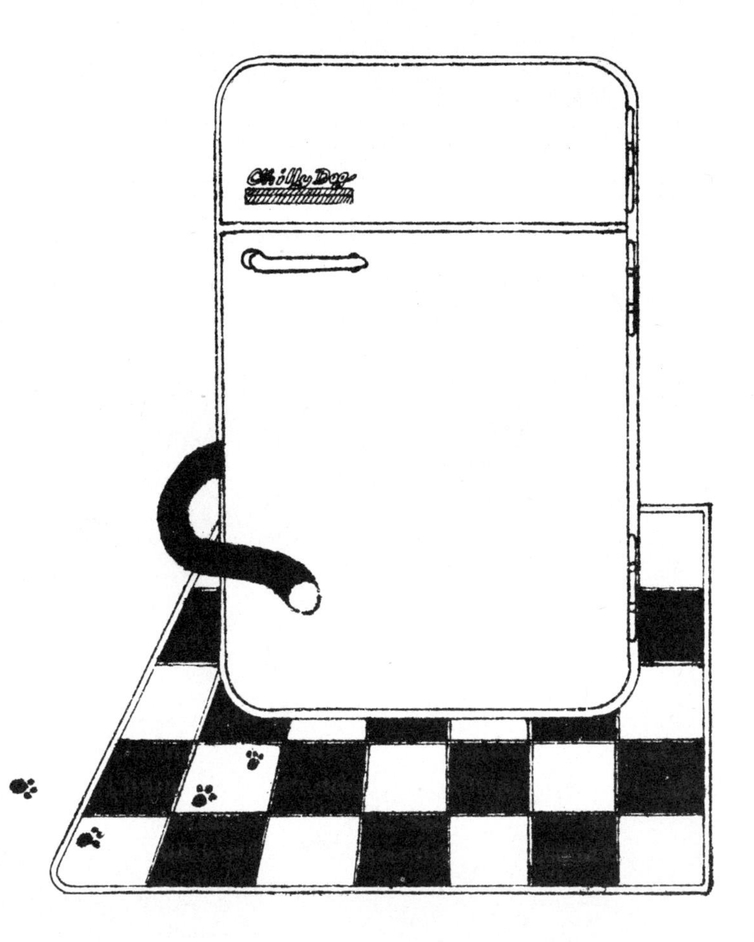

FEEDING FOR A HEALTHY PREGNANCY

Good nutrition during pregnancy and nursing is important to both the mother and the offspring. Overfed, fat animals which may also be inadequately exercised have poor muscle tone and many more difficulties with delivery than properly nourished ones. Underfed females are forced to draw upon their own body resources to nourish the growing fetuses. At best this results in a weakened bitch or queen at delivery, one who progressively can become more out of condition as nursing proceeds. At worst an unsatisfactory diet during pregnancy results in offspring which die before birth. Those puppies or kittens born alive to a malnourished mother may be at a disadvantage the rest of their lives because of inadequate nutrition during their growth in the womb. Since poor feeding during pregnancy is not often corrected during the nursing period, these puppies or kittens may miss a most important part of the feeding necessary to grow into the smart and healthy pets everyone desires.

A normal pregnancy in the dog or cat lasts approximately sixty-three days. Assuming your dog or cat is well-fed and exercised properly before breeding no special care is necessary before and during the first month of pregnancy. A normal amount of exercise is important to maintain the good muscle tone necessary for an easy delivery. Although protein supplementation is recommended during pregnancy, good quality complete and balanced commercial dog or cat foods are adequate as a basic diet during the initial stages and vitamin-mineral supplements are not necessary. If high quality proteins such as eggs, milk and meat products have not been a normal part of your pet's diet before pregnancy, offer them now but do not feed them at a level higher than 25% of the total ration.

During the second month of pregnancy a dog's or cat's calorie needs increase. Although the calorie requirement on a per pound basis increases only slightly—for dogs from about 50 to 65 calories per pound per day, for cats from about 40 to 50 calories per pound per day—the total calorie intake eventually becomes higher as pregnancy continues and the mother's weight increases due to the growth of the fetuses. A very high quality diet should be offered at this time and continued throughout lactation (nursing). Special commercial diets for pregnant and nursing pets are available or you can feed a good quality commercial high protein diet or kitten or puppy foods. Commercial diets designed for maintenance of adult pets *must* be supplemented during the last part of the pregnancy and during lactation. Use unbalanced protein supplements at no more than 20% of the diet. Balanced high quality canned meat products or orphan formulas used as dietary supplements for pregnant and nursing cats and dogs can be mixed with other foods in any proportion. Since high quality diets provide dogs and cats with all the vitamins and minerals they need, commercial vitamin-mineral supplements are not necessary. If you are very concerned about possible deficiencies, change the diet completely to a more adequate one and/or talk to a veterinarian who can advise you about supplementation that will not harm your pet or endanger the pregnancy.

It is usually impossible for a queen or bitch in the second half of pregnancy to consume all the necessary food at just one meal daily. Not only is more food needed to meet the increased calorie requirements of pregnancy, but the uterus enlarges as the fetuses grow and begins to compress the other abdominal organs. Increase the number of feedings per day in the second half of pregnancy and continue multiple feedings as lactation progresses. You can allow your dog or cat to self-feed as long as she is not becoming too fat.

Most dogs and cats restrict their exercise sufficiently as the time of delivery approaches. The last few days, however, be sure you don't encourage strenous exercise by taking your dog on long hikes or by allowing your cat to have free run of house areas where she jumps and runs. Some cats or dogs will lose their appetites for about twenty-four hours before delivery. If you are sure there are no other problems, this is no cause for concern.

Within twenty-four hours following delivery your pet should be normally interested in eating or drinking. If she is not, this calls for an examination by a veterinarian. The diet the first few days after delivery should be the same quality and approximate quantity as that fed at the end of pregnancy (about one and a third to one and one-half times the usual maintenance amount). As lactation proceeds expect your pet's food intake to increase markedly. It is impossible to say exactly how much your pet will need to eat but a rule of thumb for calorie intake to use is to feed the normal maintenance requirement *plus* 100 to 125 calories per pound

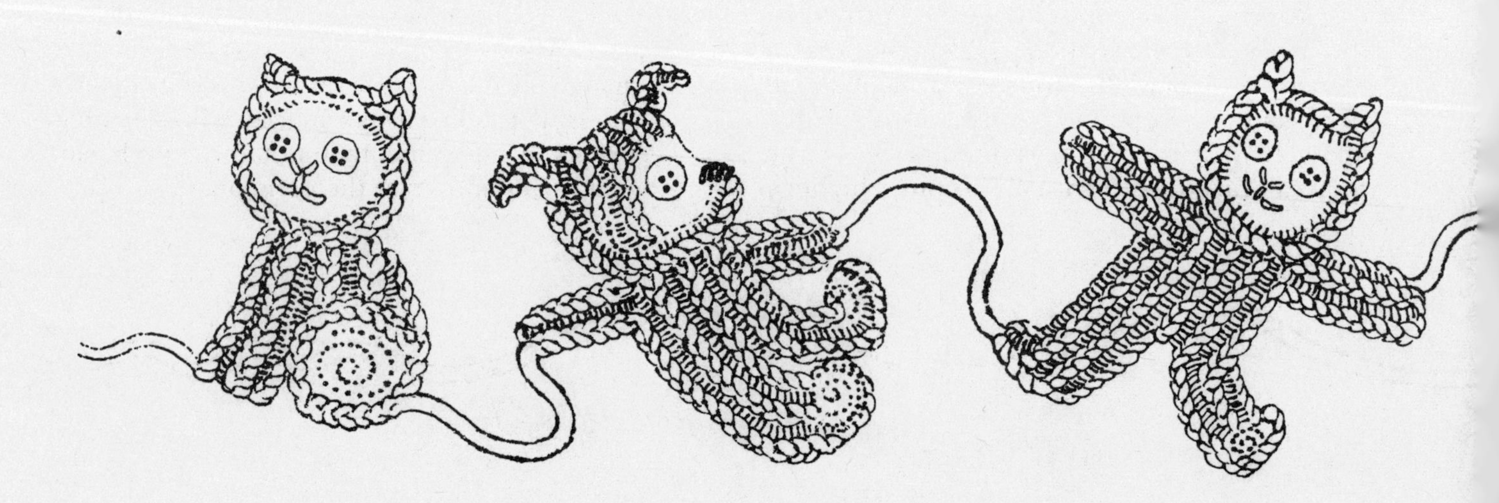

of kittens or puppies nursing. By the end of lactation a female may be consuming three or even four times as much food as she was before breeding. In fact, it is almost impossible to overfeed a nursing queen or bitch.

As in all feeding situations, use your pets appearance, behavior and body functions as guides to proper feeding during pregnancy and lactation. An abnormally thin and "worn out" looking mother may need a diet adjustment. Dogs and cats which develop diarrhea during pregnancy or lactation usually need a change in diet.

Unless you are an accomplished pet breeder, it is best to consult a good veterinarian personally about diarrhea or other digestive upsets during pregnancy or lactation. The most common cause for such problems is the failure to feed a high quality diet (one too low in digestible calories). Diarrhea is often seen when nursing mothers try to consume large quantities of dry food or very low quality canned foods to meet their high calorie needs. Dietary changes which alleviate these problems usually include decreasing or eliminating high fiber diets (some dry foods), making sure all canned products are high quality, and offering sources of easily digested calories (eggs., meat) as diet supplements to other foods. However, since health problems other than inappropriate feeding can result in digestive upsets during pregnancy or nursing, an early consultation with your pet's veterinarian can insure that her basic health is fine and that effective diet changes can be made easily.

HOW TO FEED ORPHAN KITTENS OR PUPPIES

The best care given to a mother during pregnancy does not guarantee she will raise her litter successfully. In some cases the litter is too large for the mother to care for by herself and you must serve as a foster parent. Sometimes a mother actively rejects her babies; other times delivery results in the death of the mother.

If at all possible, newborn kittens or puppies should suckle the first milk, *colostrum,* which is rich in antibodies to protect the young against disease during the first weeks of life. (If this is not possible, consult your veterinarian for further advice.) Then assume care of the orphans or foster them to another nursing mother. If fostering is attempted, try to cross foster the litter onto mothers with young ones the same size or supervise the nursing so small puppies or kittens will be sure to get sufficient milk.

Kittens and puppies which must be separated from their mothers completely must be kept in a warm environment free from drafts because they have difficulty controlling their body temperatures. From birth to about five days of age the room or box temperature should be 85° to 90°F; from about five to twenty days about 80°F. After twenty days the environmental temperature should be lowered gradually to somewhere between 70° and 75°F by the fourth week. If you don't have a human or poultry incubator, the best way to provide the proper temperature for orphan kittens or puppies is to use an electric heating pad. Hang the heating pad down one side of the box and onto about one-fourth of the bottom. Then adjust the temperature control to maintain the proper air temperature. By covering only part of the floor you allow the babies to get away from the heat if necessary. The heating pad and box bottom should be covered with cloth or newspaper which is changed each time it becomes soiled. Most authorities recommend that each kitten or puppy be kept in a separate compartment until two or three weeks old to prevent them from sucking each other's ears, tails, feet and genitals, but if they are allowed to suckle sufficiently at each nursing period, you will probably find that this is not necessary.

Orphan puppies and kittens should be fed the formula that most approaches the composition of normal mother's milk. Although you can get by with formulas made from cows' milk, cats' and dogs' milk is much higher in protein and fat and the commercial formulas are much closer to the real thing. KMR® is a well-known milk substitute for kittens. Esbilac®, Orphalac®, and Havolac® are all readily available formulas for orphan puppies and can also be used to raise orphan kittens if you cannot find a kitten milk replacer. Usually a local pet store or veterinarian will have what you need. (Commercial orphan formulas are good to supplement feed large litters and to supply extra nutrition to nursing mothers as well.)

HOME FORMULAS FOR ORPHAN PUPS OR KITTENS:

Evaporated milk . . . . .4 oz.
Water . . . . .4 oz.
Corn syrup . . . . .½ oz.
Egg yolk 1
Halibut liver oil . . . . .2 drops
Thiamine hydrochloride . . . . .1 mg

About 30 calories per ounce

Whole cow's milk . . . . .26.5 oz.
Cream (12% fat) .6.5 oz.
Egg yolk 1
Bone meal . . . . .6 gm.
Citric acid . . . . .4 gm.
Vitamin A . . . . .2,000I.U.
Vitamin D . . . . .500 I.U.

About 38 calories per ounce

Whole cow's milk .16 oz.
Corn syrup . . . . .1 tsp.
Egg yolk 1
Iodized table salt .pinch
Vitamin-mineral supplement as per pkg. or veterinarian's instructions.

About 24 calories per ounce

The best way to determine how much formula each kitten or pup needs is to weigh it and use a table of calorie requirements (see page 79). The required amount of formula is then divided into three or four portions fed at six to eight hour intervals. Small puppies and kittens seem to do better with the more frequent feedings. Example: A one half pound (225 grams) puppy needs 60 calories/pound ½ pound=30 calories per day during the first week of life. This is about one ounce formula containing thirty calories per ounce. Example: A ¼ pound kitten (112 grams) needs ¼ pound×190 calories/pound=47.5 calories per day during the first week of life. This is about 1.6 ounces of formula, a little over three tablespoonsful, containing 30 calories per ounce.

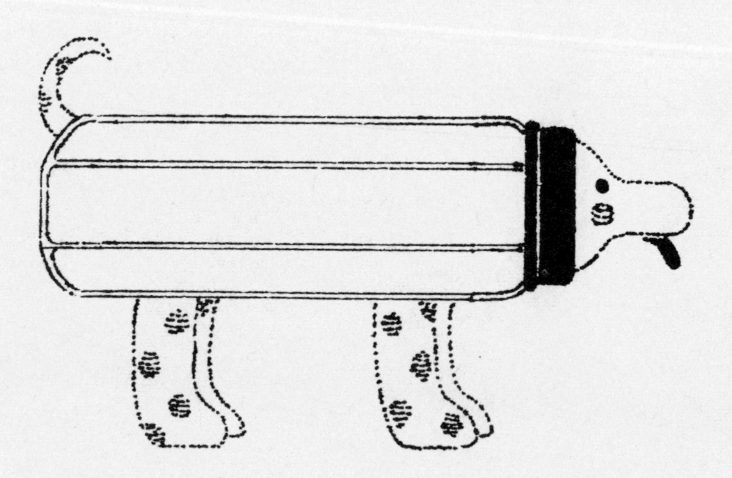

If you supply the proper caloric requirements, you do not need to feed most puppies or kittens more than three times each day. However, if the little one cannot take in the required volume at each feeding, the number of feedings must be increased. At each feeding a kitten or puppy should eat only until comfortably full—not until the abdomen is distended and tight. A steady weight gain (for kittens about 100 to 150 grams per week) and a normal stool are indications that the cat or dog is being fed properly.

All formulas are best fed after warming to body temperature (about 100°F). Formula can be administered with an eye dropper, syringe, nursing bottle or a stomach tube. A nursing bottle is usually easiest and safest in inexperienced hands. The holes in the nipple should be enlarged if the formula does not drip *slowly* from the nipple when the full bottle is inverted. For kittens and small puppies small nipples and bottles can be found at pet stores. For larger puppies, "premie" nipples or regular baby nursers and nipples can be used. Be sure the nipple size is suitable for the size puppy you are trying to feed.

Hold the kitten or puppy on its stomach. Gently separate the lips with your fingers and slip the nipple in. A healthy, hungry youngster will usually suck vigorously after tasting the milk. Use of a towel placed near the feet will give the kitten or puppy something to push and knead against as if nursing naturally. Weak babies may have to be held vertically and formula placed slowly in their mouths with an eyedropper or syringe. DO NOT place a kitten or puppy on its back to feed it or squirt liquid rapidly into its mouth. These feeding methods can cause aspiration of fluid into the lungs which can be followed by pneumonia. If you wish to use a stomach tube for feeding (the fastest method), ask your veterinarian for a demonstration.

After each feeding, the puppies or kittens should be stimulated to urinate and defecate. Moisten a cotton swab, tissue or soft cloth with warm water and gently, but vigorously, massage the ano-genital area. Nursing kittens' or puppies' stools are normally firm (not hard) and yellow. If diarrhea develops, the first thing to do is to dilute the formula by about one-half by the addition of boiled water. If this does not help within twenty-four hours, consult a veterinarian. Feeding cows' milk often causes diarrhea because of its high lactose content.

While feeding kittens or puppies artificially, keep all your equipment scrupulously clean and refrigerate or discard unused formula between feedings. Unsanitary feeding practices can produce intestinal infections which may quickly kill a young puppy or kitten.

HOW TO WEAN KITTENS OR PUPPIES

Around the age of three weeks you can begin to accustom most kittens and puppies to solid food (wean them). It is wise to begin the weaning process early and accomplish it gradually because rapid weaning often results in digestive upsets and a period of poor nourishment and impaired growth.

Place a shallow pan of formula on the floor of your pets' box or pen. (Restrain or remove the mother if she's interested in the babies' solid food.) At first the little ones will step and fall into it and make a general mess, but soon they will be lapping at the formula. When this stage is reached, meat or egg yolk blended to the texture of baby food, high protein pablem or commercial puppy or kitten foods can be added to the formula to make a gruel. After the kittens or puppies are eating the gruel, the amount of formula can be decreased gradually until they are eating solid food and drinking water from a pan left accessible at all times. Eggs, cottage cheese, yogurt and meats may all be added to a young puppy's or kitten's diet as he or she becomes adjusted to eating solid food.

Kittens and puppies with a natural mother should be allowed to continue nursing during the weaning process until they are eating well-balanced meals of solid food on their own. During this time mother dogs may sometimes regurgitate food in front of their puppies for them to consume. This is a normal part of the natural weaning process and is no cause for worry.

By five weeks of age kittens and puppies have most of their baby teeth so that the mother will usually become more and more reluctant to nurse them. As the young ones increase their intake of solid food, the mother will gradually reduce her intake of food (or you should) and she will gradually restrict the nursing time. Weaning may be acheived completely this way and this is a weaning process that comes closest to what mother nature intended. If, however, there is an actual weaning day, offer the bitch or queen water but no food or feed only a small portion of the normal maintenance diet on that day. Over the following five days gradually increase the food offered back to the normal maintenance level. This procedure helps decrease the mother's milk production.

NOTES

INDEX

(continued)

Table of Abbreviations, Weight Equivalents And Conversion Factors

WEIGHT MEASURES
1 μg = 1 microgram = 0.001 mg = 0.000001 g
1 mg = 1 milligram = 1000 μg = 0.001 g
1 g = 1 gram = 1000 mg = 0.001 kilogram
1 lb = 1 pound = 454 g = 0.454 kg
1 oz = 1 ounce = 28.4 g

LIQUID MEASURES
1 oz = 2 tablespoons = 30 milliters
1 tablespoon = 3 teaspoons = 15 milliliters

UNITS GIVEN	UNITS WANTED	FOR CONVERSION
lb	g	x 454
lb	kg	x 0.454
oz	g	x 28.4
kg	lb	x 2.2
kg	mg	x 1,000,000
kg	g	x 1000
g	mg	x 1000
g	μg	x 1,000,000
mg	μg	x 1000
oz	ml	x 30
ml	oz	x 0.033
oz	tblsp	x 2.0
tblsp	oz	x 0.5
tblsp	ml	x 15
ml	tblsp	x 0.067

IU = international unit, a measure of vitamin activity. The amount in milligrams varies depending on the vitamin under consideration.

OTHER BOOKS BY TAYLOR & NG:

WOKCRAFT by Charles & Violet Schafer. An authoritative and entertaining book on the art of Chinese wok cookery. Authentic, easy-to-follow recipes for beginners and professionals alike. Illustrated by Win Ng.

RICECRAFT. Authoress Margaret Gin delves into the fact, fiction and fancy of rice. A collection of inventive recipes takes full advantage of the international versatility of rice. Fanciful illustrations by Win Ng.

TEACRAFT—a treasury of romance, rituals, and recipes. A book on tea—its multiplicity of uses and varieties, how to test and taste, plus recipes to complement teatime. Written by Charles & Violet Schafer, illustrated by Win Ng.

BREADCRAFT by Charles & Violet Schafer. A connoisseur's collection of bread recipes: what bread is, how you make it, and how you can create your own bread style. Plus a chapter devoted to breadspreads! Illustrated by Barney Wan.

PLANTCRAFT by Janet Cox. A practical and fun guide to indoor plant care. Illustrated charts depict the growing characteristics and conditions for over 60 plant varieties. Photo gallery by L. C. Spaulding Taylor.

HERBCRAFT by Violet Schafer. The mystery of herbs unveiled: 87 pages describe 26 herbs—their origin, history, use, growing and storing conditions. Illustrated by Win Ng.

COFFEE. The story behind your morning cup: Charles & Violet Schafer elaborate on coffee—its origin, many varieties, how to brew it and what to brew it in. With recipes for companion foods. Illustrations and photography by Alan Wood.

CHINESE VILLAGE COOKBOOK. Authoress Rhoda Yee tells her story—all about the wok and wok cookery, coupled with colorful narratives on everyday life in a Chinese village. A stir-fry chart, photographic food glossary and authentic recipes guide the novice to wok mastery in no time!